Charles L. Dodson

CURRENT CHEMICAL CONCEPTS

A Series of Monographs

Editor: ERNEST M. LOEBL

R. McWeeny: Spins in Chemistry, 1970

Edited by LOUIS MEITES

Henry Taube: Electron-Transfer Reactions of Complex Ions in Solution, 1970

P. Zuman: The Elucidation of Organic Electrode Processes, 1969

Frank A. Bovey: Polymer Conformation and Configuration, 1969

H. J. Emeléus: The Chemistry of Fluorine and Its Compounds, 1969

A Polytechnic Press of the
Polytechnic Institute of Brooklyn Book

Spins in Chemistry

R. McWEENY
DEPARTMENT OF CHEMISTRY
THE UNIVERSITY
SHEFFIELD, ENGLAND

1970

ACADEMIC PRESS New York and London

ACADEMIC PRESS, INC.
111 Fifth Avenue, New York, New York 10003

United Kingdom Edition published by
ACADEMIC PRESS, INC. (LONDON) LTD.
Berkeley Square House, London W1X 6BA

Library of Congress Catalog Card Number: 72-107557

PRINTED IN THE UNITED STATES OF AMERICA

FOREWORD

This is one of a series of monographs made possible by a Science Development Grant from the National Science Foundation to the Polytechnic Institute of Brooklyn in 1965. That grant enabled the Institute's Department of Chemistry to establish a Distinguished Visiting Lectureship that was held successively by a number of eminent chemists, each of whom had played a leading part in the development of some important area of chemical research. During his term of residence at the Institute, each Lecturer gave a series of public lectures on a topic of his choice.

These monographs arose from a desire to preserve the substance of these lectures and to share them with interested chemists everywhere. They are intended to be more leisurely, more speculative, and more personal than reviews that might have been published in other ways. Each of them sets forth an outstanding chemist's own views on the past, the present, and the possible future of his field. By showing how the facts of yesterday have given rise to today's concepts, deductions, hopes, fears, and guesses, they should serve as guides to the research and thinking of tomorrow.

This volume is based on a series of six lectures and a symposium talk given by Professor Roy McWeeny while in residence at the Institute in March 1969. It is with great pride and pleasure that we present this record of the stimulation and profit that our Department, and all those able to attend the lectures, obtained from his visit.

ERNEST M. LOEBL, *Editor*
Professor of Physical Chemistry

EPHRAIM BANKS, *Acting Head*
Department of Chemistry

F. MARSHALL BERINGER
Dean of Science

PREFACE

In preparing these notes for publication I have not attempted to produce a textbook: Neither the material, nor its mode of presentation—as a series of Science Development Lectures—seemed appropriate for that purpose. The aim of the lectures, as I understood it, was to select some theme or concept of current importance in chemistry and to trace its evolution fairly systematically, from its early beginnings to its present stage of development. The first lecture should interest and entertain a large audience consisting of professors and students alike, from chemistry and from related disciplines, while the last would certainly have reached the present frontiers of research and would consequently have more appeal to the specialist. These are not the aims of a textbook. Now that I have to go into print, I feel more than ever that it would be a mistake to replace the spontaneity and informality of the lecture room by the formalities of a textbook style; the lectures are therefore published more or less exactly as they were given, with only the addition of an appendix and literature references.

The choice of topic needs some explanation. As a theoretician, I wanted to find a theme that would illustrate at some depth the deductive methods of quantum theory and their impact on many areas of chemistry. "Spin" seemed to offer the right opportunities: The idea in itself is abstruse enough to present a conceptual challenge, its assimilation into quantum mechanics beautifully illustrates the mathematical machinery of the subject, and spin has so many implications in chemistry—particularly with the rapid advance of spin resonance techniques—that no chemist can afford to be wholly ignorant of recent developments in the field. The description of spin couplings by means of a "spin Hamiltonian" goes back forty years; but even today the spin-Hamiltonian concept is widely misunderstood and frequently misused. For all these reasons, it seemed well worthwhile to go back and try to find the path leading from the principles of quantum mechanics, through what some would call the mathematical wilderness, to an understanding and interpretation of the sophisticated physical methods now employed in the investigation of molecular structure and properties.

The lectures were given during March 1969. It is a great pleasure to record my thanks to my hosts in the chemistry department, particularly to Professors Loebl, Banks, and Beringer, for their hospitality, and to all who helped make my visit such a pleasant one.

My thanks are due also to Mrs. S. P. Rogers for producing an excellent typescript from my barely legible notes, and to Academic Press and the editor of this series for their speed and efficiency in publishing this monograph.

R. McWeeny

January, 1970

CONTENTS

1

THE ORIGIN AND DEVELOPMENT OF THE SPIN CONCEPT

In this first lecture we shall discuss the historical origins of the spin concept, show how it was successfully assimilated into Schrödinger's wave mechanics, and give a preview of its implications in optical, electron spin resonance (ESR), and nuclear magnetic resonance (NMR) spectroscopy. In succeeding lectures we shall build up gradually the theory of the "spin Hamiltonian," which provides the essential link between theory and experiment.

In 1921 Stern and Gerlach set out to measure the magnetic moments of atoms by deflecting an atomic beam in an inhomogeneous magnetic field. Their experiment, in retrospect, was a milestone in physics and chemistry. First, it provided an experimental basis for the concept of spin, later introduced by Goudsmit and Uhlenbeck in their efforts to resuscitate the "old quantum theory" and reconcile it with the spectroscopic facts; second, it exemplified the "ideal measurement" in the sense of modern quantum mechanics. Indeed, it is possible to squeeze out of this one experiment a profound insight into the methods and interpretation of quantum mechanics.

The Stern–Gerlach Experiment

Let us recall the Stern–Gerlach experiment (Fig. 1.1a): Particles emerge from a furnace and a narrow beam passes between the poles of an electromagnet (shaped to produce a strongly inhomogeneous field). A magnetic dipole is deflected slightly by such a field, up or down according to the component of its moment along the field direction. When there is no field, the beam makes a spot on a photographic plate; but when the field is switched on, we find two spots, three spots, or more, depending on the atoms used (e.g., two for sodium, three for zinc). We consider the simplest case of two spots. The implication is that the component of magnetic moment can take only two values, corresponding to a magnet pointing parallel or antiparallel to the field. It is conventional to adopt the field direction as a z axis. Corresponding to the component of magnetic moment μ_z, we define an intrinsic angular momentum[1] component S_z, referred to as the "z component of spin." If the particle were actually a spinning distribution of negative charge, classical physics would predict a proportionality $\mu_z = -\beta S_z$, where the numerical constant β is the "Bohr magneton" but to allow more flexibility we write $\mu_z = -g\beta S_z$, where the numerical factor g may depend on the type of particle (e.g., for electrons it turns out that $g = 2$). This is just a hunch so far; but later it is found that S_z does possess all the properties of angular momentum, behaving just like L_z, the angular momentum due to orbital motion of a particle.

What does this experiment tell us. It indicates that S_z can be found with two possible values (λ_1 or λ_2, say, where $\lambda_2 = -\lambda_1$ since the beam is displaced equally in the up and down directions); we have made an observation that *determines* the value of

[1] It is convenient to measure all angular momenta in "dimensionless form" such that an angular momentum of L units is expressed as $\hbar L$ (where $\hbar$ is Plank's constant divided by 2π, and has the correct dimension ML^2T^{-1}).

the quantity we are trying to observe; and we find half the particles in the up-spin beam, half in the down-spin beam. If we repeat the experiment on *either one* of these beams (Fig. 1.1b), we verify that S_z has a value λ_1 (up-spin beam) or λ_2 (down-spin

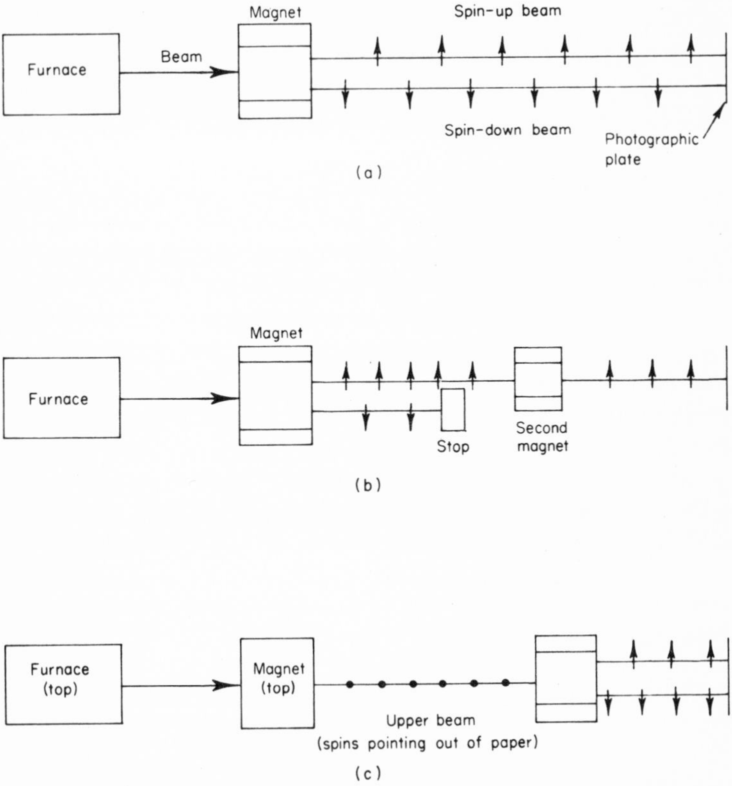

FIG. 1.1. The Stern–Gerlach experiment. (a) Spinning particles proceed from furnace (left), split by inhomogeous magnetic field. (b) Effect of a second magnet (analyzer). (c) Effect of analyzer on beam after rotation of Stern–Gerlach experiment through 90°.

beam), with no further splitting occurring. In this case the second observation does not disturb the value already recorded, and this is the criterion for an "ideal" observation in quantum mechanics. But if we twist the first analyzer around, so that the up-spins are definitely pointing toward us (Fig. 1.1c) while the down-spin beam is cut off by means of a stop, the second analyzer again splits the beam. In other words, a particle with a definite *x component* (λ_1) is found after observation to have *z component* λ_1 or λ_2, with a fifty-fifty chance of either. The incoming particle could be described as *either*

i. In a state with $S_x = \lambda_1$ *definitely*

or

ii. In a state with $S_z = \lambda_1$ (with probability $\frac{1}{2}$)
$= \lambda_2$ (with probability $\frac{1}{2}$)

Each is an equally valid description of the state of the incoming particle. If we agree to have the field in the z direction while making observations, any state of the incoming particle can be indicated statistically by giving the two fractions p_1 and p_2 into which the beam is resolved. Observation, then, consists in sorting results into categories. The result is in general indefinite and the state of a system (e.g., one particle) is described by stating the probabilities, p_1 and p_2, with which various possible results are observed.

The Formulation of Quantum Mechanics

If we could go back to 1921 and start formulating quantum mechanics again (four years before Schrödinger and Heisenberg!), we might have argued in the following way. Observation of spin always yields one of two values, λ_1 or λ_2. Let us represent states by means of an arrow of unit length (Fig. 1.2); the arrow points along the horizontal axis for the up-spin state, and along the vertical axis for the down-spin state. We call these vectors α and β and indicate any *other* state by means of another unit vector ψ whose components, c_1 and c_2 in Fig. 1.2, determine the probabilities p_1 and p_2 of getting an up-spin or down-spin result according to $p_1 = c_1^2$, $p_2 = c_2^2$. Probabilities are essentially positive numbers between 0 and 1 and this is, therefore, a sensible

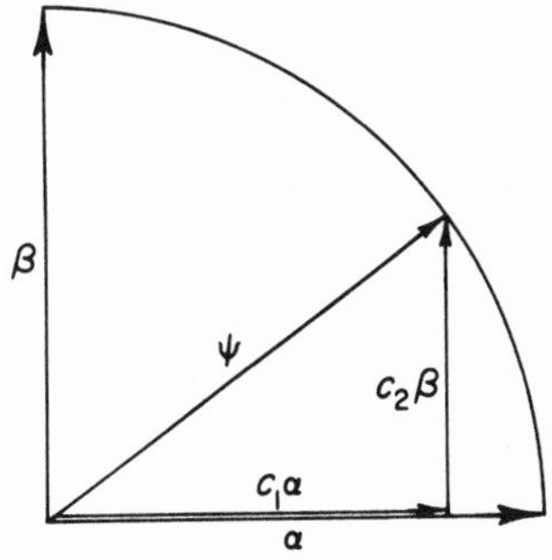

FIG. 1.2. Representation of state by a vector. α and β represent states of spin $\pm\frac{1}{2}$ while ψ is a general state (spin uncertain).

convention because $c_1^2 + c_2^2 = 1$ (square on the hypotenuse of the right triangle). The spin state of a particle is thus indicated by a *unit* vector in a two-dimensional diagram; every way the vector points indicates a different spin state. Using the triangle law for combining vectors, we write

$$\psi = c_1\alpha + c_2\beta \tag{1.1}$$

We remember also that in elementary vector theory there is a "scalar product" of two vectors[2] defined as the product of their lengths times the cosine of the angle between them. For unit vectors, $\alpha \cdot \alpha = 1$, $\beta \cdot \beta = 1$, while $\alpha \cdot \beta = 0$ (zero cosine) corresponds to perpendicular unit vectors. Also, in vector language,

$$\psi \cdot \psi = (c_1\alpha + c_2\beta) \cdot (c_1\alpha + c_2\beta) = c_1^2 + c_2^2 = 1 \tag{1.2}$$

[2] Indicated by putting a dot between them.

expresses the unit length of ψ, which conventionally represents an arbitrary spin state.

In the experiment just described, particles in state ψ are knocked into states with vectors α or β, although for a given particle we could not say which, as a result of the interaction between particle and observer. With the *operation* of sorting out the up-spin particles, we can associate the "projection" of ψ onto α. A "projection operator" P_1 picks out the α part of ψ; thus

$$\mathsf{P}_1\psi = c_1\alpha \tag{1.3}$$

In a similar way, for selection of down-spin particles, we might write

$$\mathsf{P}_2\psi = c_2\beta \tag{1.4}$$

If we form the scalar products $\psi \cdot \mathsf{P}_1\psi$ and $\psi \cdot \mathsf{P}_2\psi$, we obtain

$$\begin{aligned} \psi \cdot \mathsf{P}_1\psi &= (c_1\alpha + c_2\beta) \cdot c_1\alpha = c_1^2(\alpha \cdot \alpha) = c_1^2 = p_1 \\ \psi \cdot \mathsf{P}_2\psi &= (c_1\alpha + c_2\beta) \cdot c_2\beta = c_2^2(\beta \cdot \beta) = c_2^2 = p_2 \end{aligned} \tag{1.5}$$

And if (back in 1921) we had been able to make the imaginative leap that characterizes all great discoveries, this observation might have led us to associate *operators* with the quantities we measure, anticipating the developments that, in fact, took place a few years later. If we had made a large number of spin observations and represented the spin state by ψ (for any prepared state of the incoming particles), we would have noticed that the average or "expectation value" of S_z could be written [using (1.5)]

$$\langle S_z \rangle = p_1\lambda_1 + p_2\lambda_2 = \psi \cdot (\lambda_1\mathsf{P}_1 + \lambda_2\mathsf{P}_2)\,\psi = \psi \cdot \mathsf{S}_z\psi \tag{1.6}$$

where

$$\mathsf{S}_z = (\lambda_1\mathsf{P}_1 + \lambda_2\mathsf{P}_2) \tag{1.7}$$

is an *operator* associated with the observable S_z.

States and observables may now be "represented" in terms of a vector diagram. The operator S_z defined in (1.7) does something to any vector ψ representing a state (Fig. 1.3): P_1 yields the part lying along the horizontal axis, λ_1 makes this part λ_1 times as long; $\lambda_2\mathsf{P}_2$ has a similar effect on the vertical component; and if we add the parts together to get a new vector and then take the scalar product with ψ, as in (1.6), we find a number, which is the average or expectation value of S_z, that we should obtain after a long series of measurements. To verify that this prescription works, we go back to the situation in Fig. 1.1c: In this case $p_1 = p_2 = \frac{1}{2}$ and the state was represented by a vector ψ at 45° to α and β; but the vectors ψ and $\mathsf{S}_z\psi$ in Fig. 1.3 are then perpendicular, giving the correct expectation value $\langle S_z \rangle = \psi \cdot \mathsf{S}_z\psi = 0$ (equal probability of positive and negative components, zero average).

We are laying the foundations of quantum mechanics! You can play the same game, perhaps in a more familiar context, with two Kekulé structures: The very essence of quantum mechanics is

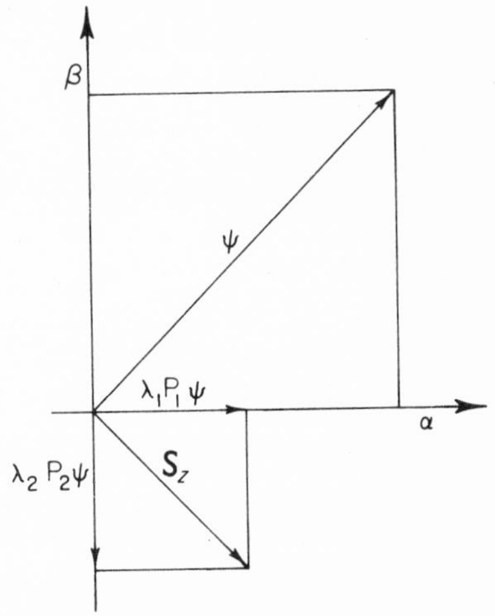

FIG. 1.3. Vector diagram representing the effect of the spin operator $\mathsf{S}_z = \lambda_1\mathsf{P}_1 + \lambda_2\mathsf{P}_2$ (with $\lambda_1 = \frac{1}{2}$, $\lambda_2 = -\frac{1}{2}$).

the representation of a state in which something is "uncertain" (in this case, the allocation of the double bonds) as a "mixture" or "superposition" of states in which that something is definite (e.g., the allocation of bonds in a Kekulé structure). Of course, in discussing the structure of quantum mechanis and the nature of its operators we must start at the beginning; and for that reason we begin with the most primitive kind of system: a single spin.

Expectation Values and Eigenvalues

The expectation value is usually written with the bracket notation due to Dirac

$$\langle S_z \rangle = \langle \psi \mid \mathsf{S}_z \mid \psi \rangle \tag{1.6'}$$

a "bra" $\langle \psi \mid$ and a "ket" $\mid \psi \rangle$, with the operator sandwiched between. Operators are associated with all the quantities we measure, and the value we expect to find in any given state is always given by the same formula, although the vector may not be in two dimensions. For example, the vector components c_r, with r taking values 1 and 2 only, may be replaced by a component $c(x)$ taking a continuous range of values as the label x runs over an infinite number of points. In this case the state may be represented directly by a function $\psi(x)$; an operator A is any mathematical prescription yielding a new function $\psi'(x) = \mathsf{A}\psi(x)$; and the scalar product or bracket quantity, giving the expected value of the quantity with which A is associated, is defined as

$$\langle A \rangle = \langle \psi \mid \mathsf{A} \mid \psi \rangle = \int \psi^*(x)\, \mathsf{A}\psi(x)\, dx \tag{1.8}$$

a formula that will be familiar to everyone as one of the basic postulates of quantum mechanics.

As these lectures are concerned mainly with the quantum mechanics of spin systems, it is worth developing the necessary mathematical machinery just a little bit further in this particular context. For those of you who are relative newcomers to quantum mechanics, this might help to remove some of the conceptual difficulties; for those of you who are experts, it might be an entertaining digression.

The spin operator S_z normally produces from any state vector ψ a new vector ψ' pointing in a different direction; but there are two special cases in which it does not. If ψ points along the horizontal axis, $\psi = \alpha$, then P_2 projects out nothing (no vertical component), while P_1 leaves α unchanged; thus $\mathsf{S}_z\alpha = \lambda_1\alpha$. Similarly, if $\psi = \beta$, we obtain $\mathsf{S}_z\beta = \lambda_2\beta$. These two states, however, are the only ones in which measurement of S_z is *certain* to yield one and only one definite value, $S_z = \lambda_1$ for $\psi = \alpha$ or $S_z = \lambda_2$ for $\psi = \beta$. The equation

$$\mathsf{S}_z\psi = \lambda\psi \tag{1.9}$$

(i.e., S_z multiplies the length of the vector ψ by a numerical factor λ) thus provides a *test* of whether S_z definitely has one of its possible values (λ_1 or λ_2) in the state represented by ψ, i.e., whether the beam of particles is "pure" with all spins up or all down. There are only *two* solutions: $\psi = \alpha$, satisfying the equation when λ has the numerical value λ_1, and $\psi = \beta$ when $\lambda = \lambda_2$. The solutions are called *eigenvectors* (ψ) and *eigenvalues* (λ), the latter giving the "quantized values," $S_z = \lambda_1$ or λ_2, that observation of the spin z component can yield. Everyone will recognize that this "test for purity" is what we are using whenever we write down Schrödinger's equation

$$\mathsf{H}\Psi = E\Psi \tag{1.10}$$

Here the operator H is associated with the energy of the system (we consider not just the spin, but rather the energy associated with motion through space of all the constituent particles), Ψ is a "wave function," and the values of E for which solutions can be found are the permitted energy levels of the system. The equation itself is called an *eigenvalue equation*, the solutions being *eigenfunctions* and *eigenvalues*.

Properties of the Spin Operators

All we need now, to get through the whole of quantum chemistry, are the properties of the operators associated with the things we are going to observe. It is not always possible to get these by very satisfactory arguments; we must appeal to the idea that Newtonian dynamics must come out as a limiting case when we deal with macroscopic rather than atomic systems. But for spin, which is what really concerns us at the moment, we can do better.

Let us therefore talk specifically about an electron and give the eigenvalues λ_1 and λ_2 $(= -\lambda_1)$ their observed values, which (in units of $\hbar$, see p. 2) are $+\frac{1}{2}$ and $-\frac{1}{2}$. Then

$$\mathsf{S}_z\alpha = \tfrac{1}{2}\,\alpha, \qquad \mathsf{S}_z\beta = -\,\tfrac{1}{2}\,\beta \tag{1.11}$$

which means S_z has the effect of multiplying α by $\frac{1}{2}$ (or β by $-\frac{1}{2}$) without changing its length. Thus, $\mathsf{S}_z^2 = \mathsf{S}_z\mathsf{S}_z$ (S_z applied twice in succession) would multiply α by $\frac{1}{4}$ and would clearly do the same to β since $(-\frac{1}{2})(-\frac{1}{2}) = \frac{1}{4}$. Consequently, $\mathsf{S}_z^2\alpha = \frac{1}{4}\alpha$, $\mathsf{S}_z^2\beta = \frac{1}{4}\beta$, and

$$\mathsf{S}_z^2\psi = \tfrac{1}{4}\,\psi \tag{1.12}$$

since both α and β components of $\psi = c_1\alpha + c_2\beta$ are multiplied by $\frac{1}{4}$.

In other words, S_z^2 is equivalent to multiplication by a number: We might write $\mathsf{S}_z^2 = \frac{1}{4}\mathsf{I}$ where the "identity operator" I leaves any vector in the "spin space" unchanged and the numerical factor just reduces the length of the vector to $\frac{1}{4}$ of its original value.

To find some other properties of the operators, we can remark that space is "isotropic," i.e., the same can be said for the x and y

directions as for the z direction, and hence, that S_x^2 and S_y^2 must also merely multiply by $\frac{1}{4}$. Thus, the squared magnitude of the spin angular momentum should be $\frac{3}{4}$, since

$$\mathbf{S}^2\psi = (S_x^2 + S_y^2 + S_z^2)\,\psi = \tfrac{3}{4}\,\psi$$

We write this result

$$\mathbf{S}^2\psi = S(S+1)\,\psi$$

with the spin quantum number S taking the value $\frac{1}{2}$, and note that although the largest allowed *component* of spin is $\frac{1}{2}$, the magnitude of the vector is of the form $[S(S+1)]^{1/2}$, which you will recall is a characteristic and classically inexplicable property of angular momentum in general. Now if we consider an ***arbitrary*** direction, the cosines of the angles made with the axes being l, m, and n (with $l^2 + m^2 + n^2 = 1$), any component along the new direction (and hence, any expectation value) must be related to those for the x, y, and z directions by (using a prime to denote the new axis)

$$\langle S_z' \rangle = l\langle S_x \rangle + m\langle S_y \rangle + n\langle S_z \rangle$$

The corresponding new operator, referring to the z component along the new direction in space, is thus $S_z' = lS_x + mS_y + nS_z$; since space is isotropic, this must have the same properties as any other component. But

$$\begin{aligned} \mathbf{S}_z'^2 &= l^2S_x^2 + m^2S_y^2 + n^2S_z^2 + lm(S_xS_y + S_xS_y) + \cdots \\ &= \tfrac{1}{4}\mathrm{I} + lm(S_xS_y + S_yS_x) + \cdots \end{aligned}$$

and if this is equivalent to $\frac{1}{4}\mathrm{I}$ for arbitrary values of l, m, and n, we must have

$$(S_xS_y + S_yS_x) = (S_yS_z + S_zS_y) = (S_zS_x + S_xS_z) = 0 \qquad (1.13)$$

Thus, if we apply S_yS_x to any vector representing a spin state, then S_xS_y, then add the results, we have nothing left. The spin operators are said to "anticommute." The argument now almost goes by itself: To characterize the operators completely, we need to find what they do to spin states α and β, for which we already know that $S_z\alpha = \frac{1}{2}\alpha$, $S_z\beta = -\frac{1}{2}\beta$. Simple algebraic manipulation shows that the relations

$$\begin{aligned} S_x\alpha &= \tfrac{1}{2}\beta, \qquad & S_y\alpha &= \tfrac{1}{2}i\beta, \qquad & S_z\alpha &= \tfrac{1}{2}\alpha \\ S_x\beta &= \tfrac{1}{2}\alpha, \qquad & S_y\beta &= -\tfrac{1}{2}i\alpha, \qquad & S_z\beta &= -\tfrac{1}{2}\beta \end{aligned} \tag{1.14}$$

complete the specification of a vector representation of spin properties. These equations form the basis of the *Pauli* theory of spin, which provides an adequate framework for all present day applications in molecular theory. Most of you are no doubt familiar with the properties of orbital angular momentum operators (L_x, L_y, L_z). You will be able to prove easily from (1.14) that the spin operators must satisfy similar *commutation relations*

$$\begin{aligned} S_xS_y - S_yS_x &= iS_z \\ S_yS_z - S_zS_y &= iS_x \\ S_zS_x - S_xS_z &= iS_y \end{aligned} \tag{1.15}$$

which are characteristic of any kind of angular momentum. The commutation relations for spin and orbital angular momentum provide a cornerstone in the theory of atomic spectra and, indeed, in many areas of theoretical physics. They show, for instance, that *many*-particle systems will be characterized by a *resultant* spin of magnitude $[S(S+1)]^{1/2}$, with S a half-integer and with z component taking quantized values $S, S-1, \ldots, -S$ (integer steps). The rules characterizing the combination of spins apply equally well to the coupling of all kinds of angular momenta.

We now have all the mathematical background for getting spin into chemistry. But let us pause for a moment and remember where it came from: one experiment, a little speculation, and a few general ideas about vectors and the isotropy of space. Of course, this is not the way things happen in history; but the possibility of constructing so much from so little, with elegance and economy, seems to be one of the greatest challenges—and sources of intellectual pleasure—in theoretical work.

Spin has implications of two kinds: those implications which arise merely from the need to recognize it in classifying states, and those which depend on small interactions involving the associated magnetic dipole. In the first category, we think immediately of Pauli's exclusion principle, or its wave mechanical equivalent that "two electrons in the same orbital must have opposite spins," which provides the ultimate basis for the periodic table. In the second category, we think of the various magnetic resonance techniques at present sweeping through all parts of chemistry. The remaining lectures will be devoted to both areas; here we merely want to introduce some of the fundamental notions.

Spin and Antisymmetry

In the absence of spin, Schrödinger's equation for a one-electron system reads

$$\mathsf{h}\phi = \epsilon\phi \tag{1.16}$$

where h is the usual one-electron Hamiltonian (a simple differential operator) and ϕ is an *orbital* depending on the spatial variables ($\mathbf{r}$, say) describing the position of the electron in space. The solutions $\phi_1, \phi_2, \ldots,$ occur for certain specific values of the parameter ϵ ($\epsilon_1, \epsilon_2, \ldots,$ say) which are the *orbital energies*. Everyone knows about the 1s, 2s, 2p,... orbitals for an electron moving about a central nucleus, these being the atomic orbitals, which are building bricks in valence theory. The existence of spin means it is not sufficient to describe a state by means of an orbital, *even when we neglect small spin terms in the Hamiltonian.* We must now use a *spin-orbital* from the set

$$\phi_1\alpha, \quad \phi_1\beta, \quad \phi_2\alpha, \quad \phi_2\beta, \ldots, \phi_n\alpha, \quad \phi_n\beta, \ldots$$

The reason for the product form is simply that this represents a state in which *both* characteristics (orbital energy and spin component) are definite; thus, the eigenvalue test shows that

$$\mathsf{h}(\phi\alpha) = \epsilon(\phi\alpha) \qquad \text{energy definite}$$

$$\mathsf{S}_z(\phi\alpha) = \tfrac{1}{2}(\phi\alpha) \qquad \text{spin component definite}$$

since h works only on the factor ϕ and S_z only on α. The states available to an electron moving around a nucleus are thus

$$1s\alpha, \quad 1s\beta, \quad 2s\alpha, \quad 2s\beta, \quad \ldots$$

For a many-electron system, we classify states in a similar way. Thus, if we denote the available spin-orbitals for any atom or molecule by

$$\psi_A = A\alpha, \quad \bar{\psi}_A = A\beta, \quad \psi_B = B\alpha, \quad \bar{\psi}_B = B\beta, \quad \ldots$$

then, with neglect of electron interaction, a spin-orbital *product* such as

$$\psi_A(1)\,\bar{\psi}_A(2)\,\psi_B(3)\cdots$$

(where the integers indicate which electrons are referred to) is an eigenfunction of all the operators $\mathsf{h}(1)$, $\mathsf{S}_z(1)$, $\mathsf{S}_z(2)$,..., and hence represents a state in which electron 1 is in A with spin $+\frac{1}{2}$, electron 2 is simultaneously in A with spin $-\frac{1}{2}$, and so on. Each electron has "its own" energy, and, in this crude approximation, the total energy would be the sum $E = \epsilon_A + \epsilon_A + \epsilon_B + \cdots$. Taking over the Bohr–Sommerfeld "aufbau" approach to atomic structure, electrons are assigned to spin-orbitals, and the lowest energy (i.e., the ground state) goes with an assignment to the lowest energy spin-orbitals. But why is there no "collapse" with all electrons going to the 1s orbital?

The first remarkable consequence of the introduction of spin was the resolution of this dilemma. An electronic wave function has the statistical interpretation that $|\Psi(1, 2,...)|^2$ determines the probability of finding the first electron at point 1, the second simultaneously at point 2, etc. Indistinguishability of the particles requires that $|\Psi(2, 1,...)|^2$ should have the same numerical value. Hence, $\Psi(2, 1,...) = \pm\Psi(1, 2,...)$, i.e., the electronic wave function is *either* symmetric *or* antisymmetric.[3] Experiment decides unequivocally in favor of antisymmetry for electrons, and the immediate consequence is that a state which assigns two electrons to the same orbital with opposite spins is represented by

$$\Psi_a(1, 2) = \psi_A(1)\,\bar{\psi}_A(2) - \bar{\psi}_A(1)\,\psi_A(2)$$

(changing sign for the interchange of 1 and 2), while two electrons in the same orbital with the *same* spin would give

$$\Psi_s(1, 2) = \psi_A(1)\,\psi_A(2) - \psi_A(1)\,\psi_A(2) = 0$$

It is impossible, in fact, to construct orbital wave functions with the required antisymmetry unless the electrons all occupy *different spin-orbitals*. The *ad hoc* postulates used in the Bohr–Sommerfeld model are replaced by an all-embracing *antisymmetry principle*. The configurational theory of atomic structure, with the gradual filling of "shells" as one proceeds through the periodic table, is too well known to require further comment; but we should not lose sight of its roots in the spin and symmetry properties of electrons, or of the fact that the new principle applies universally, permitting us to discuss molecules and crystals merely by changing from atomic orbitals to molecular orbitals or crystal orbitals.

[3] It can be shown that there are only two distinct possibilities and that these are mutually exclusive.

Spin and Magnetism

The one-electron Hamiltonian referred to so far is derived by associating suitable operators with each of the observables (the kinetic and potential energies) in the classical expression for the energy of an electron. For a many-electron system, it is a sum of such contributions, supplemented by interaction terms. We can compute energy levels using such a Hamiltonian and obtain a reasonable account of the main features of, e.g., atomic spectra. Sometimes the situation is particularly simple, as in sodium where there is one electron outside an inert gas "core" and spectral series are associated with changes of orbital of this "series" electron. Thus, the sodium d lines are associated with an electron jumping between a 3s and a 3p orbital; but why are there *two* lines, very close together? And why do these lines acquire a more complicated fine structure when a strong magnetic field is applied? Such effects are in fact due to the presence of the spin magnetic moment.

Let us apply a strong magnetic field and consider the effect on a 3s electron. There is no orbital angular momentum, but nevertheless, this level is resolved into a doublet. Evidently the energies of the two possible spin-orbitals, $3s\alpha$ and $3s\beta$, are slightly different. We recall that the components of magnetic moment in the two states are $\mu_z = -g\beta S_z$ and that, for a classical dipole, the potential energy in a field of strength B (along the z axis) would be $-B\mu_z$. The equivalent spin term in the Hamiltonian operator would thus be $g\beta B\mathsf{S}_z$ and if we take

$$\mathsf{H} = \mathsf{h} + g\beta B\mathsf{S}_z \tag{1.17}$$

and remember that $\mathsf{h}\phi_{3s} = \epsilon_{3s}\phi_{3s}$ for the orbital part of the wave function we find at once that

$$\begin{aligned} \mathsf{H}(\phi_{3s}\alpha) &= (\epsilon_{3s} + \tfrac{1}{2}\, g\beta B)\,(\phi_{3s}\alpha) \\ \mathsf{H}(\phi_{3s}\beta) &= (\epsilon_{3s} - \tfrac{1}{2}\, g\beta B)\,(\phi_{3s}\beta) \end{aligned} \tag{1.18}$$

The level is split into two, the separation of the branches being $g\beta B$, both states still being eigenfunctions of the more complete Hamiltonian. The observed Zeeman effect thus finds an elementary explanation.

Turning to the 3p state, we must now take account of the *orbital* angular momentum: Even before an external field is applied, the electron sees the nucleus as a rapidly moving, positively charged particle and experiences a resultant nonzero magnetic field. This suggests that, even in the absence of an applied field, there will be an "internal Zeeman effect." This is exactly what happens. With no external field, the $\pm\frac{1}{2}$ spin component sets along the axis of orbital angular momentum, and the spin-orbit coupling splits the 3p level into two branches. The 3s–3p transition then leads to the yellow doublet in the sodium spectrum. In a p state the orbital angular momentum is $l = 1$ and the coupled states correspond to a *resultant* $j = 1 \pm \frac{1}{2} = \frac{3}{2}$ or $\frac{1}{2}$. If, then, we finally apply an external magnetic field, each j-state gives a number of states corresponding to its own total angular momentum, two for $j = \frac{1}{2}$ (just like a free spin) but four for $j = \frac{3}{2}$, corresponding to z components going in unit steps from $\frac{3}{2}$ down to $-\frac{3}{2}$.

The picture we now have, with magnetic interactions included, is as shown in Fig. 1.4, and we can now turn to some of the experimental implications of the fine structure. If we have sodium atoms in, say, a discharge tube there will be many atoms in each available state, i.e., all states will be populated. Different kinds of external perturbation will then induce different kinds of transition.

In general, large quanta of radiation in the visible or ultraviolet range will induce transitions between widely spaced levels: For the sodium atom the 3s—3p separation is about 16,500 cm^{-1} (or about 2.1 eV) and the two jumps indicated by the left-hand arrows in Fig. 1.4 give rise to the yellow D lines in the optical spectrum. The Zeeman splitting, on applying a magnetic field, then arises as indicated on the right (not to scale!).

The smaller quanta associated with radiation in the microwave region may induce transitions between different Zeeman levels of each individual group (on the right in Fig. 1.4), whose separation may be about 28,000 MHz ($\sim$1 cm^{-1}) for an electron

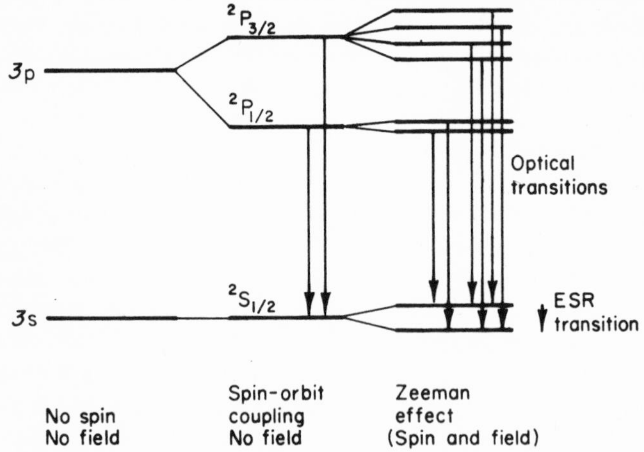

FIG. 1.4. Energy levels for the sodium "series" electron (schematic). Optical transitions (before and after applying a magnetic field) are indicated by the long arrows. The short arrow (extreme right) indicates an ESR transition. The scale has been greatly increased to show the fine structure.

in a field of 10,000 Gauss. It is these latter transitions within the fine structure of a single electronic level that interest us in electron spin resonance (ESR) experiments and which we shall consider in a later lecture; they involve a change of magnetic dipole component and are therefore induced by an oscillating magnetic field, usually transverse to the main static field which fixes the axis of quantization. The principles of such an experiment are indicated in Fig. 1.5 together with an observed resonance line.[4] Since there must be a resultant spin[5] before resonance can be observed, ESR experiments give valuable information about free radicals, ions, and molecules in triplet states.

Finally, in very high resolution spectroscopy, other effects come to light. There is one kind of spin we have still not considered: The nuclei themselves are many-particle systems and frequently possess a resultant *nuclear spin*. So far we have considered only the "fine structure" of electronic levels; but we have noticed that a single ESR line *still* exhibits structure (Fig. 1.5). When nuclear spins are admitted, we must go one more order of magnitude finer and look at this "hyperfine structure" due to coupling between electronic and nuclear effects. Hyperfine splittings are commonly of the order of 10^{-3} cm^{-1} corresponding to quanta in the radio frequency region (42 MHz for a proton at 10,000 Gauss). Their origin is indicated in Fig. 1.6, which shows us why, in ESR experiments at high resolution, a resonance peak shows a structure which, with several nuclei, may become very complicated. In *nuclear magnetic resonance* (NMR) experiments there is no change of electronic state, but a quantum of radio frequency energy is absorbed and the nuclear spin state changes. The change in coupling energy then yields important information about the electronic environment of the nuclei. Clearly NMR is a useful tool for investigating states with zero *electron* spin (i.e., the majority of molecules in

[4] The horizontal scale is usually in Gauss because it is easier in practice to find the resonances by varying the static field B than the frequency of the oscillating field; a frequency difference is related to a field difference by $h\,\Delta\nu = \mu_z\,\Delta B$.

[5] It is of course clear that the observed angular momentum (in either a Stern–Gerlach or any other kind of experiment) is a *resultant*, arising from the coupling of orbital and spin angular momenta of all the individual electrons. Frequently, for one electron outside a closed shell, as in the case of the sodium atom, the "resultant" is effectively the angular momentum of the single electron; and for molecules of low symmetry this angular momentum is due to *spin only*, the orbital contribution being "quenched."

their ground states), as long as some of the nuclei present possess nonzero spins; and since protons have spin $\frac{1}{2}$, NMR is of major importance in organic chemistry. Figure 1.7, for example, shows how the position of a proton resonance peak varies very slightly according to chemical environment and how structural information can be obtained from NMR spectra in this way.

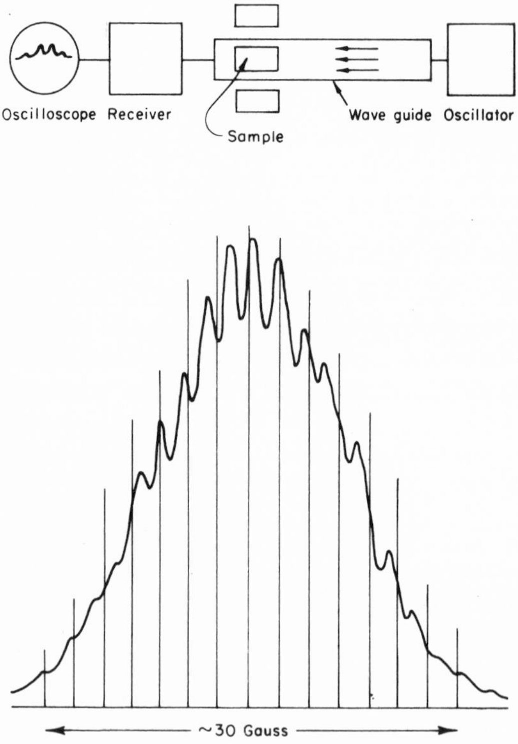

FIG. 1.5. Principle of ESR experiment, and a typical resonance peak from an organic free radical. The "hyperfine structure" is due to interaction with nuclear spins.

A preliminary discussion can take us no further. We know how to describe spin states, how to handle spin operators, and how both can be incorporated in the fabric of quantum

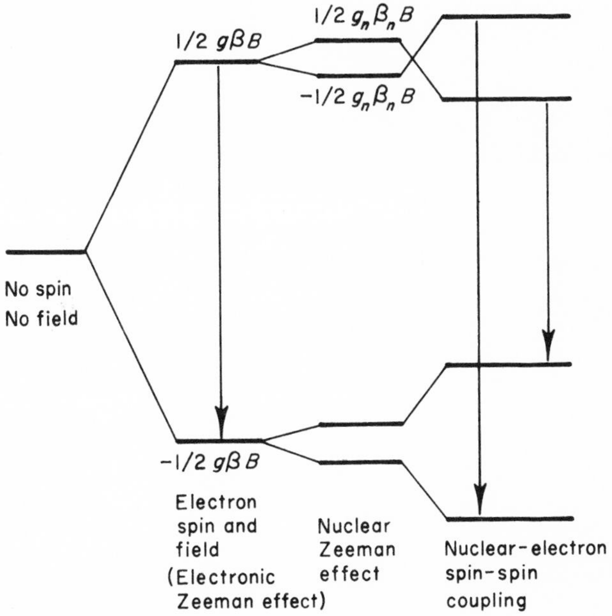

FIG. 1.6. Hyperfine splitting of ESR transitions (long arrows) for one electron in the presence of one proton. The electronic Zeeman splitting (left) is modified, each level being raised or lowered by a nuclear energy term, and then further disturbed by electron-nuclear interaction. The resultant peak is split into two (cf. Fig. 1.5, where the structure is more complex) by the hyperfine coupling.

mechanics. Indeed, spin systems provide some of the simplest possible illustrations of quantum mechanical principles and methods. We are now in a position to follow the ramifications of

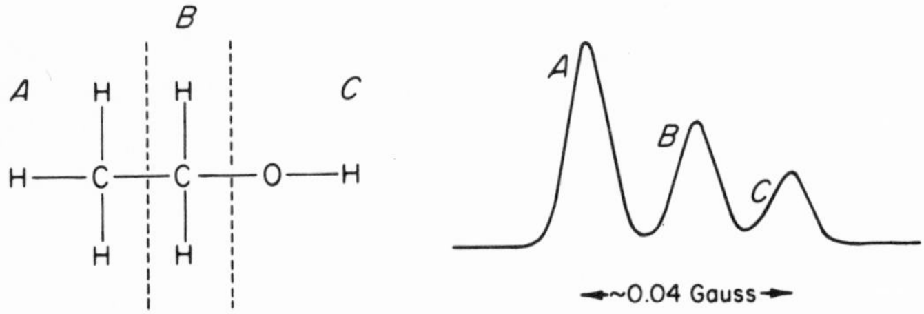

FIG. 1.7. NMR peaks, showing dependence of signals on chemical environment. The three peaks arise from the three types of proton—those in regions *A*, *B*, and *C*, respectively. Areas under the peaks indicate the numbers of protons of each kind.

the spin concept into many areas of chemistry, first into the interpretation of the covalent bond and then into the elucidation of magnetic effects.

2

SPIN AND VALENCE

For many years, following Heitler and London's classic paper on the chemical bond in the hydrogen molecule, discussions of valence were dominated by ideas concerning the "coupling" of electron spins. Chemistry was provided with a graphic and picturesque language that was adopted with great enthusiasm; covalent bonds came to be associated with the "pairing," or "antiparallel coupling," of the spins of electrons on different atoms, and it was supposed that an important factor in breaking bonds and many other processes was the energy needed to "uncouple" the spins. Heitler, London, Van Vleck, and the other pioneers of quantum chemistry appreciated, of course, that *there are no strong coupling forces* between electron spins and that spin properties are merely related (through the Pauli, or anti-symmetry, principle) to the spatial location of electron pairs; but in the excitement of the early days, caution was thrown to the winds. It seemed that a few general principles could account for the basic properties of the chemical bond in all its variety, that the precise arithmetical details would anyway remain inaccessible, and that what really mattered was the construction of a comprehensive, if somewhat formal, general theory. The result was that an elaborate *semi*theoretical approach to valence problems—embodied in "valence bond" or "resonance" theory with all its terminology and manifold applications—was to reign supreme in many areas of chemistry for fifteen or twenty years. It therefore seems important that before discussing the spin interactions that really exist (i.e., the small magnetic effects associated with spin angular momentum) we should look at the ones that really do *not* exist.

Let us go back to the Heitler–London calculation. The two hydrogen atoms, at a large distance, would each have an electron in a 1s orbital, i.e., electronic wave functions $a(\mathbf{r}_1)$ and $b(\mathbf{r}_2)$, say, where $\mathbf{r}_1$ and $\mathbf{r}_2$ stand for the position variables of electrons 1 and 2. As always, a wave function for the systems taken together, but with interaction neglected, would be the product $a(\mathbf{r}_1)\, b(\mathbf{r}_2)$. We know from Lecture 1 that the spin description must be included by adding spin factors and that antisymmetry must be imposed by making suitable permutations and combining the results with appropriate signs. Before proceeding in more detail, however, we must develop the notation a little.

We know that an orbital, a say, is a function of position in space (position vector $\mathbf{r}$) and that its squared value at $\mathbf{r}$, $|a(\mathbf{r})|^2$ $[=a^*(\mathbf{r})a(\mathbf{r})]$, indicates the probability of finding the electron there.[1] The spin factor we are going to add is usually written formally in the same way as a function of a "spin variable" s, which we may imagine to be the spin component $S_z(=s)$ along the axis of quantization (conventionally taken as the z axis). This function is defined at only two points; thus, keeping the same kind of statistical interpretation as for an orbital function, the spin function $\alpha(s)$ vanishes unless $s = \frac{1}{2}$, meaning that the spin component has zero probability of being different from $\frac{1}{2}$ (i.e., half an "atomic unit") in state α; similarly $\beta(s)$ vanishes unless $s = -\frac{1}{2}$, corresponding to the certainty of finding spin $-\frac{1}{2}$ in state β. In fact α and β, which we first introduced as unit

[1] $|a(\mathbf{r})|^2$ is actually the probability *per unit volume* of finding the electron in the vicinity of point $\mathbf{r}$.

vectors (Lecture 1) in a two-dimensional "spin space" are written as functions merely for formal convenience; for this way of writing them allows us to handle spin factors just like orbitals, without having to use Pauli matrices or summations over indices repeatedly. You can visualize the spin functions α and β as "spikes" in the vicinity of $s = +\frac{1}{2}$ and $s = -\frac{1}{2}$, respectively, and they have to be normalized just like orbitals to permit the usual statistical interpretation. Thus, the unit vector and orthogonality properties of the vectors representing the two spin states, namely $\langle\alpha \mid \alpha\rangle = \langle\beta \mid \beta\rangle = 1$ and $\langle\alpha \mid \beta\rangle = \langle\beta \mid \alpha\rangle = 0$, are translated into

$$\int \alpha^*(s)\,\alpha(s)\,ds = \int \beta^*(s)\,\beta(s)\,ds = 1$$

and (2.1)

$$\int \alpha^*(s)\,\beta(s)\,ds = \int \beta^*(s)\,\alpha(s)\,ds = 0$$

In wave-function language the normalization equation for $\alpha(s)$, for example, states that the probability of finding *some value* of the spin, obtained by "summing" over all s values, is unity. The spike is, therefore, infinitely narrow but infinitely high, the limit being approached in such a way that the integral converges to the value 1. Such a function, the Dirac delta function, is "improper," and whole books have been written about its strict mathematical interpretation and justification; but ever since the nineteenth century it has been widely used, in one form or another, in applied mathematics.

Now that we understand how to represent wave functions describing both space and spin [e.g., in terms of spin-orbital products such as $a(\mathbf{r}_1)\alpha(s_1)b(\mathbf{r}_2)\beta(s_2)$] we can return to the Heitler–London calculations. The results of setting up antisymmetric wave functions, with one electron in *a* and one in *b*, are well known.[2] Two appropriate functions are, using $\mathbf{x}$ to stand for the space and spin variables $(\mathbf{r}, s)$ collectively,

$$\Psi_g(\mathbf{x}_1, \mathbf{x}_2) = N_g[a(\mathbf{r}_1)\, b(\mathbf{r}_2) + b(\mathbf{r}_1)\, a(\mathbf{r}_2)]\, [\alpha(s_1)\, \beta(s_2) - \beta(s_1)\, \alpha(s_2)]$$

$$\Psi_u(\mathbf{x}_1, \mathbf{x}_2) = N_u[a(\mathbf{r}_1)\, b(\mathbf{r}_2) - b(\mathbf{r}_1)\, a(\mathbf{r}_2)]\, [\alpha(s_1)\, \beta(s_1) + \beta(s_1)\, \alpha(s_2)] \tag{2.2}$$

where N_g and N_u are normalizing factors. The first function is a "singlet," the second a "triplet"; the other two functions of the triplet have spin factors $\alpha(s_1)\, \alpha(s_2)$ and $\beta(s_1)\, \beta(s_2)$ but the same orbital factor. If we evaluate the energy expectation values, using

$$E = \langle \Psi \mid \mathsf{H} \mid \Psi \rangle \tag{2.3}$$

with the usual spinless Hamiltonian,

$$\mathsf{H} = \sum_i \mathsf{h}(i) + \tfrac{1}{2} \sum_{i,j} g(i, j) \tag{2.4}$$

where $g(i, j) = e/r_{ij}$ is an electron interaction and

$$\mathsf{h}(i) = -(\hbar^2/2m)\, \nabla^2(i) + V(i) \tag{2.5}$$

is the Hamiltonian for each electron i moving in the field of the nuclei (with potential energy V), we find two different energies because the *orbital* factors differ; but orbital and spin factors are related through the antisymmetry principle and it is clear what Van Vleck [*2*] meant by calling the spin coupling an "indicator." The energy formula for the two cases (upper and lower signs, respectively) is usually written in the form

$$E = (Q \pm K)/(1 \pm \Delta) \tag{2.6}$$

where

$$Q = \langle ab \mid \mathsf{H} \mid ab \rangle$$
$$K = \langle ab \mid \mathsf{H} \mid ba \rangle \tag{2.7}$$
$$\Delta = \langle ab \mid ba \rangle = S_{ab}^2$$

[2] See, e.g., Pauling and Wilson [*1*], p. 340 *et seq.*

(S_{ab} being the overlap integral between the orbitals a and b). The quantity Q is called the Coulomb integral because it contains, in addition to the "personal" energies of the electrons in their own orbitals, an interaction energy representing the classical Coulomb repulsion of two charge distributions, with densities $|a|^2$ and $|b|^2$, just as if each electron were smeared out. The "exchange" integral K is often described, by contrast, as a "nonclassical" term. Provided overlap is not too large, we may neglect Δ and, on adding the repulsive energy of the two nuclei V_{nn}, write

$$E_{\text{tot}} \simeq Q + V_{nn} \pm K \tag{2.8}$$

for the total energy of the system. The behavior of E_{tot} for singlet and triplet is very familiar (Fig. 2.1). About nine-tenths of the binding energy comes from the K term and, therefore, the bond came to be associated with "exchange energy."

To introduce the spin explicitly we need to know what

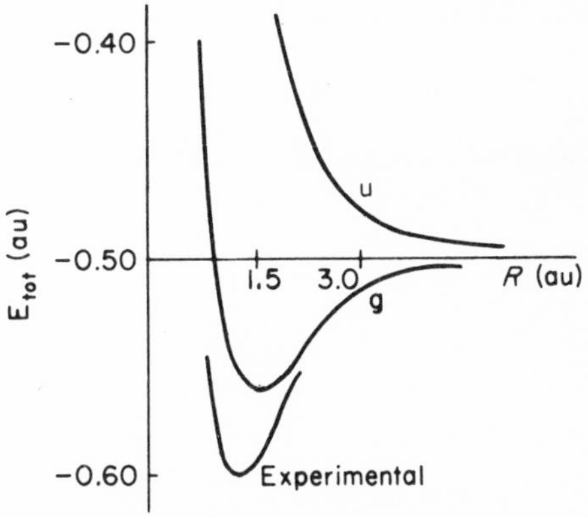

FIG. 2.1. Results of the Heitler–London calculation. Curves g and u give total energies as a function of internuclear distance (R) for the lowest singlet (gerade) and triplet (ungerade) states, compared with the experimental ground-state curve.

situations the alternative spin factors describe. We recall the properties of the spin operators, given in (1.14):

$$\begin{aligned} \mathsf{S}_x\alpha &= \tfrac{1}{2}\beta, & \mathsf{S}_y\alpha &= \tfrac{1}{2}i\beta, & \mathsf{S}_z\alpha &= \tfrac{1}{2}\alpha \\ \mathsf{S}_x\beta &= \tfrac{1}{2}\alpha, & \mathsf{S}_y\beta &= -\tfrac{1}{2}i\alpha, & \mathsf{S}_z\beta &= -\tfrac{1}{2}\beta \end{aligned} \tag{2.9}$$

for operators and spin functions referring to any electron. If we have two electrons, we can define a *total* spin vector, whose components are the sums of corresponding components of the individual electrons. The observable values of these components are given in terms of the total spin operators

$$\mathsf{S}_x = \mathsf{S}_x(1) + \mathsf{S}_x(2), \quad \mathsf{S}_y = \mathsf{S}_y(1) + \mathsf{S}_y(2), \quad \mathsf{S}_z = \mathsf{S}_z(1) + \mathsf{S}_z(2)$$

and the observable values of the squared magnitude of total spin are given in terms of[3]

$$\mathbf{S}^2 = \mathsf{S}_x^{\,2} + \mathsf{S}_y^{\,2} + \mathsf{S}_z^{\,2}$$

This may be written in the rearranged form

$$\mathbf{S}^2 = \mathbf{S}^2(1) + \mathbf{S}^2(2) + 2\mathbf{S}(1)\cdot\mathbf{S}(2)$$

where $\mathbf{S}^2(i)$ is analogous to $\mathbf{S}^2$ but refers only to electron i, while $\mathbf{S}(1)\cdot\mathbf{S}(2)$ denotes the "scalar product operator"

$$\mathbf{S}(1)\cdot\mathbf{S}(2) = \mathsf{S}_x(1)\,\mathsf{S}_x(2) + \mathsf{S}_y(1)\,\mathsf{S}_y(2) + \mathsf{S}_z(1)\,\mathsf{S}_z(2) \tag{2.10}$$

By testing the effect of the operator $\mathbf{S}(1)\cdot\mathbf{S}(2)$ on all four products of two spin functions, we obtain a remarkable result due to Dirac [*3*], namely that

$$\mathbf{S}(1)\cdot\mathbf{S}(2) = \tfrac{1}{4}(2\mathsf{P}_{12} - 1) \tag{2.11}$$

[3] The quantity on the left is to be interpreted as a single operator (the one associated with the square of the spin angular momentum) rather than the square of an operator.

where P_{12} interchanges the spin variables s_1, s_2. This equivalence holds in the sense that the two operators (on the left and right, respectively) have exactly the same effect on any spin product such as $\alpha(s_1)\,\beta(s_2)$ and hence on any spin function constructed from the various possible products. This spin-exchange identity, widely used by Dirac and Van Vleck, was a cornerstone in the development of valence bond theory.

As a first application, we cast the Heitler–London result in a new form. If Θ is any one of the spin functions in the four hydrogen molecule states, we obtain

$$[\mathbf{S}(1) \cdot \mathbf{S}(2)]\,\Theta = -\tfrac{3}{4}\,\Theta \qquad \text{(singlet)} \qquad \text{or} \qquad \tfrac{1}{4}\,\Theta \qquad \text{(triplet)}$$

Since we already know that $\mathbf{S}^2(i) = \frac{3}{4}$, it follows that $\mathbf{S}^2\Theta = [\frac{3}{4} + \frac{3}{4} - 2(\frac{3}{4})]\Theta$ for a singlet while $\mathbf{S}^2\Theta = [\frac{3}{4} + \frac{3}{4} + 2(\frac{1}{4})]\Theta$ for a triplet; and thus $\mathbf{S}^2\Theta = S(S+1)\Theta$, where the resultant spin quantum number is $S = 0$ (singlet) or $S = 1$ (triplet). In other words, the antisymmetric and symmetric spin functions describe eigenstates of *total* spin.

The energy values in the singlet and triplet states may now be correctly reproduced by a "spin only" formula. We note that

$$\langle \Theta \mid \mathbf{S}(1) \cdot \mathbf{S}(2) \mid \Theta \rangle = -\tfrac{3}{4} \qquad \text{(singlet)} \qquad \text{or} \qquad \tfrac{1}{4} \qquad \text{(triplet)}$$

and hence obtain (cf. Equation (2.6) with neglect of Δ)

$$E = \langle \Theta \mid \mathsf{H}_s \mid \Theta \rangle \tag{2.12}$$

where

$$\mathsf{H}_s = Q - \tfrac{1}{2}K[1 + 4\mathbf{S}(1) \cdot \mathbf{S}(2)] \tag{2.13}$$

This is not the electronic Hamiltonian of the hydrogen molecule; it is a *fictitious* Hamiltonian containing only spin operators and numerical parameters, whose expectation values in singlet and triplet spin states coincide with those of the real Hamiltonian in states represented by corresponding approximate wave functions.[4] This was probably the first "spin Hamiltonian" ever used in a molecular context, although Dirac and Van Vleck had already used this technique for obtaining neat expressions for the energies of states arising from certain atomic configurations. It provides an attractive formal prescription for "absorbing" the complexities of a detailed energy calculation into a few numerical parameters: There is no actual physical coupling between the spins (neglecting the small magnetic effects) but nevertheless the energy formula makes it look as if there were! The use of spin Hamiltonians in many areas of chemical physics is now commonplace, but we must always remember they provide the means of *fitting* observed results and identifying parameter values within a certain theoretical scheme—they do not provide a "first principles" method of calculating anything.

Valence Bond Theory

In view of the resurgence of interest in valence bond (VB) theory (particularly in the interpretation of ESR and NMR experiments) it is worth looking briefly back to its beginnings. Kekulé put forward his graphical schemes for representing chemical structures in 1859; but it was not until 1929 that the physical meaning of valence bond diagrams became clear. Today, we start from a one-configuration wave function—an allocation of electrons to orbitals (ignoring any *doubly* occupied orbitals, whose electrons are not available for interatomic spin coupling)—and consider all ways of describing by means of spin functions the potential couplings of valence-electron spins. The analog of the Heitler–London function is then

$$\Phi = (N!)^{-1/2}\,\mathsf{A}[\Omega(\mathbf{r}_1, \mathbf{r}_2, \ldots, \mathbf{r}_N)\,\Theta(s_1, s_2, \ldots, s_N)] \qquad (2.14)$$

where $\Omega(\mathbf{r}_1, \mathbf{r}_2, \ldots, \mathbf{r}_N) = \phi_1(\mathbf{r}_1)\,\phi_2(\mathbf{r}_2) \cdots \phi_N(\mathbf{r}_N)$ is an orbital

[4] Subject to neglect of the overlap term Δ.

product, Θ describes some intuitively suitable spin-coupling scheme, and the antisymmetrizer A imposes antisymmetry, in accordance with the Pauli principle, by permuting variables in all possible ways and summing the results with appropriate signs. The orbitals are assumed orthonormal, and in that case the factor in front of the A gives a correctly normalized wave function. By analogy with the hydrogen molecule calculation, it is natural to assume that

$$\Theta_1(s_1, s_2, \ldots, s_N) = \theta(s_1, s_2)\, \theta(s_3, s_4) \cdots \qquad (2.15)$$

in which

$$\theta(s_i, s_j) = |\, \alpha(s_i)\, \beta(s_j) - \beta(s_i)\, \alpha(s_j) \,|/\sqrt{2}$$

will yield a wave function describing bonding between orbitals ϕ_1 and ϕ_2 (orbitals holding electrons with paired spins), orbitals ϕ_3 and ϕ_4, etc. But an alternative spin-coupling scheme would be

$$\Theta_2(s_1, s_2, \ldots, s_N) = \theta(s_1, s_3)\, \theta(s_2, s_4) \cdots$$

and there would obviously be many more. Who is to say which is most appropriate? This is simply another way of putting Kekulé's question: How should we draw the chemical bonds so as to satisfy the "valences" of a number of atoms? Quantum mechanics has the answer: We set up functions $\Phi_1, \Phi_2, \ldots$, corresponding to spin factors $\Theta_1, \Theta_2, \ldots$, and approximate the wave function by a mixture

$$\Psi = c_1\Phi_1 + c_2\Phi_2 + \cdots + c_\kappa\Phi_\kappa + \cdots \qquad (2.16)$$

The coefficients, whose squares yield the "statistical weights" of the various bonding schemes, are then mathematically determined by the *variation method* which yields a best approximation to the lowest energy (i.e., ground state) solution of the Schrödinger equation. Just as the spin state of an individual particle in the Stern–Gerlach experiment is described by a mixture $\psi = c_1\alpha + c_2\beta$ of different possibilties, so the valence bond allocation in a molecule is described as a mixture of all possible situations (cf. pp. 7–8). The different bond allocations, as we all know, are represented pictorially in the diagrams or "structures" first drawn by Kekulé; thus

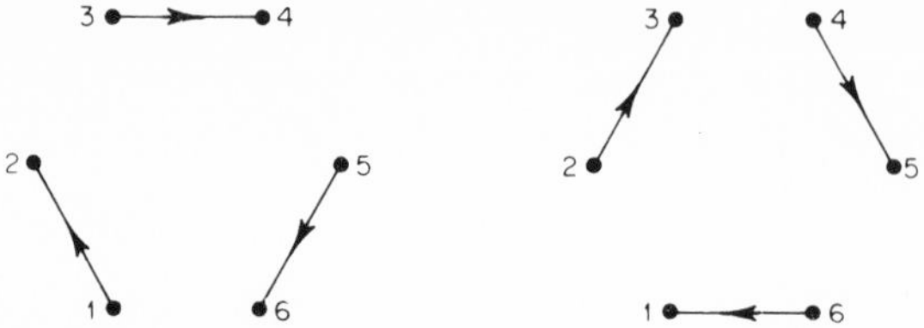

would be alternative bonding schemes for the six electrons providing the benzene double bonds, and would serve equally well to indicate two spin-coupled functions. The actual variational procedure to determine the coefficients c_1, c_2,..., and hence the relative importance of the different bonding schemes, is to solve the "secular equations"

$$\begin{aligned} (H_{11} - EM_{11})\, c_1 + (H_{12} - EM_{12})\, c_2 + \cdots &= 0 \\ (H_{21} - EM_{21})\, c_1 + (H_{12} - EM_{12})\, c_2 + \cdots &= 0 \\ \cdots \end{aligned} \qquad (2.17)$$

in which the matrix elements $H_{\lambda\kappa}$ and nonorthogonality integrals $M_{\lambda\kappa}$ are given by

$$H_{\lambda\kappa} = \langle \Phi_\lambda \mid \mathsf{H} \mid \Phi_\kappa \rangle \qquad M_{\lambda\kappa} = \langle \Phi_\lambda \mid \Phi_\kappa \rangle \qquad (2.18)$$

and can be evaluated for any assumed set of structures.

Some aspects of the quantum mechanical description, in particular the use of "spin-paired" functions, were in fact anticipated in a remarkable paper by Sylvester [*4*] (then professor of mathematics at Johns Hopkins University) in 1878! He was working on the theory of invariants and was intrigued by Kekulé's paper. In the more leisurely days of nineteenth century science, it was evidently possible for a scientist to look beyond the boundaries of his own discipline and derive pleasure and satisfaction from work in distant fields in a way that seems hardly possible any longer.

His paper was entitled "On an application of the new atomic theory to the graphical representation of the invariants and covariants of binary quantics" and explains that "by the new atomic theory I mean that sublime invention of Kekulé ... ." He continues:

> Casting about, as I lay awake in bed one night, to discover some means of conveying an intelligible conception of the objects of modern algebra to a mixed society mainly composed of physicists, chemists and biologists, interspersed only with a few mathematicians, to which I stood engaged to give some account of my recent researches in this subject of my predilection, and impressed as I had long been with the feeling of affinity, if not identity of object, between the inquiry into compound radicals and the search for "Grundformen" or irreducible invariants, I was agreeably surprised to find of a sudden distinctly pictured on my mental retina a chemicographical image serving to embody and illustrate the relations of these derived algebraic forms to their primitives and to each other, which would perfectly accomplish the object I had in view, as I will now proceed to explain.

We offer an explanation which can differ little, except in interpretation (and a return to a less extravagent scientific prose), from Sylvester's. Each factor such as $\theta_{12}(s_1, s_2)$ is *invariant* under a rotation of axes of spin quantization in the sense that if $\alpha \to \alpha' = a\alpha + b\beta$ and $\beta \to \beta' = c\alpha + d\beta$ then[5]

$$\alpha'(s_1)\,\beta'(s_2) - \beta'(s_1)\,\alpha'(s_2) = \alpha(s_1)\,\beta(s_2) - \beta(s_1)\,\alpha(s_2)$$

The wave function is unchanged under rotations, and it is this invariance that expresses the zero value of total spin, since the spin components (vector components) would be affected by

[5] From the requirement that α' and β' are normalized and orthogonal, as are α and β.

change of axes if any one of them were nonzero. All the spin-paired functions associated with Kekulé type diagrams are similarly invariant (every factor invariant) and therefore describe singlet states (zero *total* spin) of the whole many-electron system. This construction of spin eigenfunctions is the simplest one we can possibly imagine; it can be extended to the case $S \neq 0$ by allowing, say, n α-factors, in addition to the pairs, in which case $S = \frac{1}{2}n$ and all functions of the same n are "covariants" (i.e., transform in a similar way under rotations). It is now known [*5*, *6*] that the Kekulé-type structures (using dots for the "free" spins) can be related easily to the standard "branching diagram" functions set up by a more laborious mathematical procedure [*7*].

To give an example, five spins may be coupled to a particular resultant $S = \frac{1}{2}$ in several ways. Each way corresponds to a "path" in the branching diagram Fig. 2.2a, leading from the origin to the endpoint with $S = \frac{1}{2}$. The possible paths are indicated in Fig. 2.2b, the first corresponding to a function like (2.15) with spins 1 and 2 paired, 3 and 4 paired, and 5 left free; the corresponding pairing schemes (Kekulé diagrams) are shown in Fig. 2.2c. The functions of type (2.15) (Sylvester's covariants) are not identical with the brancing diagram functions, but the latter may be obtained by orthogonalizing each of the functions in the last line against those that precede it (going from left to right).

More than fifty years after Sylvester, in 1932, Rumer, Teller, and Weyl [*8*] unwittingly and quite independently reversed his argument (i.e., using Kekulé diagrams to explain the theory of invariants) by writing a paper called "A basis of binary vector invariants, suitable for the valence theory" in which they applied invariant theory to the manipulation of Kekulé structures.

There are certain difficulties in developing Kekulé's attractive picture as a rigorous quantum mechanical method of calculating molecular wave functions. For the moment, we ignore them and proceed in a formal way. If the orbitals ϕ_1, ϕ_2,..., ϕ_N are assumed orthogonal (as in neglecting Δ in the Heitler–London calculation) it is easy to obtain matrix element rules, obtaining the elements $H_{\lambda\mu}$, in (2.18) for any pair of structures from the "superposition pattern" arising from their Kekulé–Rumer diagrams. Each matrix element takes the form $Q + \sum_{i<j} a_{ij}K_{ij}$ where the

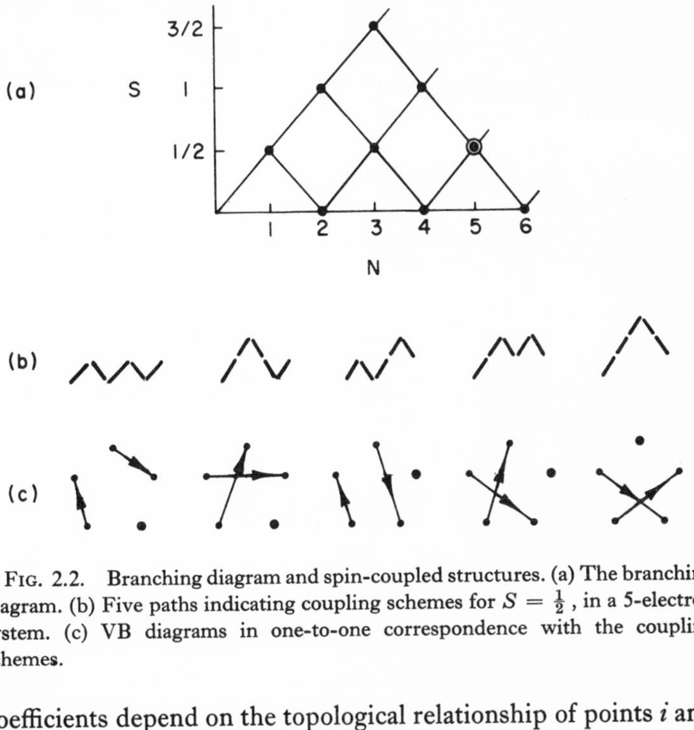

FIG. 2.2. Branching diagram and spin-coupled structures. (a) The branching diagram. (b) Five paths indicating coupling schemes for $S = \frac{1}{2}$, in a 5-electron system. (c) VB diagrams in one-to-one correspondence with the coupling schemes.

coefficients depend on the topological relationship of points i and j in the pattern; and if the secular problem (2.17) is then solved, the ground state energy finally appears in the form

$$E = Q + \sum_{ij} b_{ij}K_{ij} \tag{2.19}$$

where the coefficients b_{ij} depend on the numerically determined weights c_1, c_2,..., with which the different structures occur in the wave function expansion. Q and K_{ij} are defined like corresponding quantities in the Heitler–London calculation

$$Q = \langle \phi_1\phi_2 \cdots \phi_N \mid \mathsf{H} \mid \phi_1\phi_2 \cdots \phi_N \rangle$$

$$K_{ij} = \langle \phi_1 \cdots \phi_i \cdots \phi_j \cdots \phi_N \mid \mathsf{H} \mid \phi_1 \cdots \phi_j \cdots \phi_i \cdots \phi_N \rangle = \langle \phi_i\phi_j \mid g \mid \phi_j\phi_i \rangle$$

where $g = g(i, j)$ is the electron interaction term e^2/r_{ij}, and the reduction of K_{ij} depends on the assumed orthogonality of the atomic orbitals.

The importance of the spin-coupling scheme emerges when we go over to a spin-Hamiltonian formulation. If we write the wave function

$$\Psi = \mathsf{A}(\Omega\Theta)/\sqrt{N!} \tag{2.20}$$

where

$$\Theta = c_1\Theta_1 + c_2\Theta_2 + \cdots \tag{2.21}$$

is any normalized combination of the spin eigenfunctions, the expectation energy value is (for a spinless Hamiltonian) $E = \langle \mathsf{A}(\Omega\Theta) | \mathsf{H} | \mathsf{A}(\Omega\Theta) \rangle / N!$ and reduces easily to

$$E = \sum_{\mathsf{P}} \epsilon_{\mathsf{P}} \langle \Omega \mid \mathsf{H} \mid \mathsf{P}^{(r)}\Omega \rangle \langle \Theta \mid \mathsf{P}^{(s)}\Theta \rangle \tag{2.22}$$

where the summation is over all permutations, acting on both space and spin variables ($\mathsf{P}^{(r)}$ and $\mathsf{P}^{(s)}$, respectively). With orthogonal orbitals, permutations other than single interchanges give vanishing contributions, and so we find

$$E = Q - \sum_{ij} \langle \Theta \mid \mathsf{P}_{ij}^{(s)} \mid \Theta \rangle K_{ij} \tag{2.23}$$

Finally, from the Dirac identity (2.11) the energy may be rewritten as the expectation value of a *spin Hamiltonian*

$$\mathsf{H}_s = Q - \tfrac{1}{2} \sum_{i<j} K_{ij}[1 + 4\mathbf{S}(i) \cdot \mathbf{S}(j)] \tag{2.24}$$

in a pure spin state Θ. Thus

$$E = \langle \Theta \mid \mathsf{H}_S \mid \Theta \rangle \tag{2.25}$$

which reduces exactly to the Heitler–London result in the two-electron case.

The expectation values of the spin terms in H_S, namely

$$\langle \Theta \mid -\tfrac{1}{2}[1 + 4\mathbf{S}(i) \cdot \mathbf{S}(j)] \mid \Theta \rangle$$

for the various choices of i and j, may now be identified with the coefficients b_{ij} in the energy expression (2.19) obtained by solving the secular problem; in other words, from the energy expression we can obtain the expectation value of the spin scalar product for the electrons associated with any "bond" i—j involving orbitals ϕ_i and ϕ_j. We say that $\langle \mathbf{S}(i) \cdot \mathbf{S}(j) \rangle = -\frac{3}{4}$ for paired spins and $\frac{1}{4}$ for parallel spins. If Θ is any *single* structure and i, j are in *different* pairs, corresponding to "random" coupling between spins, we find a third value $\langle \mathbf{S}(i) \cdot \mathbf{S}(j) \rangle = 0$. On taking the expectation value of H_S and putting in these results, we obtain the *perfect pairing formula*

$$E = Q + \sum_{\substack{i<j \\ \text{paired}}} K_{ij} - \sum_{\substack{i<j \\ \text{parallel}}} K_{ij} - \sum_{\substack{i<j \\ \text{random}}} K_{ij} \tag{2.26}$$

For many years this formula was widely used in qualitative discussions of the shape and stability of polyatomic molecules, the interpretation of empirical additivity rules, and even in the discussion of energy surfaces for chemical reactions.

For states which are *not* well described by a single structure (i.e., where "resonance" among several alternatives is appreciable) the preceding observations also form the basis of a "bond order" definition due to Penney [*9*] and developed by Moffitt [*10*]: We simply take

$$p_{ij} = \langle -\tfrac{4}{3}\, \mathbf{S}(i) \cdot \mathbf{S}(j) \rangle \tag{2.27}$$

since, whenever *one* Kekulé–Rumer structure adequately

describes the bonding, linkage of positions i and j gives $p_{ij} = 1$ while nonlinkage gives $p_{ij} = 0$. This definition has certain advantages over the more arbitrary definition (weighted mean of "bonds" and "no bonds") used by Pauling and others in discussing the π electrons of conjugated molecules. It has acquired a new importance recently in the theory of the transmission of nuclear spin-spin coupling, (e.g., Barfield [*11*]; see also, Karplus and Anderson [*12*]).

Difficulties and Developments

The preceding results are formally satisfactory, and there is no denying that the valence bond theory in the form due to Heitler, London, Rumer, Weyl and developed further by Slater, Pauling, Eyring, and others is an extremely elegant and flexible theory—all the difficulties being absorbed into the parameters or, if one puts it more unkindly, "swept under the carpet." When the age of nonempirical calculations arrived, these difficulties came to light and there was a general swing to molecular orbital (MO) theory. What went wrong with valence bond theory? What ideas can we usefully salvage from it?

First, it should be emphasized that, provided the orbitals are suitably chosen, a valence bond wave function may often describe a molecular state with fairly high precision, sometimes considerably better than a simple molecular orbital function (as in the case of the hydrogen molecule calculation). Work by Matsen and Browne [*13*] has given excellent functions for LiH with fewer independent functions than in a corresponding MO calculation. But the crux of the matter is the choice of the orbitals, which seem to have disappeared from the analysis because we became preoccupied with the spin coupling and the introduction of a spin Hamiltonian. In fact, the orbitals that give the best wave functions, of spin-paired form, are heavily *non*-orthogonal, and this is where our difficulties really begin.

At first sight, it seems strange that overlap should be so important, since in the simplest case $E = (Q + K)/(1 + S^2)$, and the S^2 seems to make little difference, the binding energy coming from K. But when $S \neq 0$ we find (real orbitals)

$$K = \langle ab \mid \mathsf{H} \mid ba \rangle = \langle ab \mid g \mid ab \rangle + 2S\langle a \mid \mathsf{h} \mid b \rangle \qquad (2.28)$$

where only the first term (which is essentially positive) is a genuine exchange integral; this is outweighed by the second term, proportional to overlap, which makes K negative. The bond, therefore, really has nothing to do with exchange of electrons; the $\langle a \mid \mathsf{h} \mid b \rangle$ term contains energy of attraction between the positive nuclei and an *ab* "overlap term" in the electron distribution, and we are thrown back on a Hellmann–Feynman "classical" interpretation, with the buildup of charge between the nuclei being the important feature of the bond. Thus, in the traditional development of valence bond theory, we accept a dangerous inconsistency: keeping overlap terms when it suits us but rejecting them when it does not. This is basically the reason why the theory, in its original form, cannot be employed in nonempirical calculations, as was first pointed out by Slater [*14*].

There are several ways out of these difficulties:

i. We may retain overlap integrals and do a "first principles" calculation, writing the spin eigenfunctions in terms of Slater determinants and using the nonorthogonal generalization of Slater's rules [*15*]. This is extremely laborious (except for few-electron systems such as those dealt with by Matsen and Browne [*13*]), and general VB-type matrix element formulas have never been obtained in a convenient closed form.

ii. We may start from an orthogonalized set of orbitals, *making* all overlap integrals vanish rigorously. In this case, valence bond structures of the type considered so far give no bonding whatever, and to get the bonds back we have to admit extensive "configuration interaction" with the "polar" structures [*16a–c*] which arise when electrons are transferred between atoms to yield orbital configurations in which some AO's are doubly occupied (negative) and others are empty (positive).

iii. We may make progress in particular cases, e.g., when a single structure describing a unique set of bonds is appropriate, by using a completely different approach in which the individuality of the bonds is emphasized and the wave function is written as an antisymmetrized product of functions for the separate electron pairs, possibly with a function of Heitler–London type for each bond [*17–20*].

All these avenues have been explored. The first is practicable for small molecules and gives good wave functions, as we have seen from the work already cited. The second is generally applicable (with polar structures included), but practical difficulties arise from the large number of structures to be handled. These difficulties are, however, no worse than those experienced in any other type of extended configuration interaction calculation; thus Fig. 2.3 gives an early example of the convergence of MO and VB calculations on benzene, toward a common end point. The third possibility proves to be an attractive and generally applicable method in all situations where the "perfect pairing approximation" is effective (i.e., for saturated molecules with well-localized bonds) or, more generally, when distinct "groups" of electrons (not necessarily pairs) may be recognized [*19*].

For large molecules consisting of recognizably distinct parts, in relatively weak interaction, this last approach holds out considerable hope of progress at a nonempirical level. For a set of electron pairs, a single structure is replaced by a wave function of the form

$$\Phi = M\mathsf{A}[\Phi_A(1, 2)\, \Phi_B(3, 4)\, \Phi_C(5, 6) \cdots] \tag{2.29}$$

in which the *different* pairs are described by functions built up from mutually orthogonal sets of orbitals. The energy expression then takes the perfectly rigorous form (applicable even when the pairs are replaced by many-electron groups)

$$E = \sum_{R} E^{R} + \sum_{R<S} (J^{RS} - K^{RS}) \tag{2.30}$$

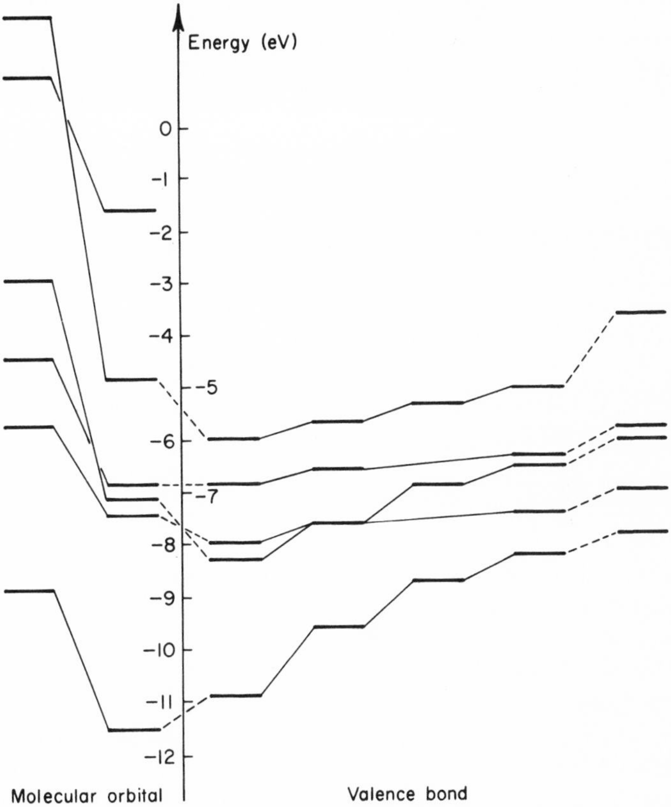

FIG. 2.3. Convergence of MO and VB calculations for benzene π-electron states [*16c*] MO results are for one configuration (extreme left) and up to nine configurations. VB results (read from right to left) with up to 89 structures.

where J^{RS} and K^{RS} are generalized forms of the Coulomb and exchange integrals, referring to interactions between *whole groups* (not just orbitals). Thus, J^{RS} is the Coulomb repulsion between the electrons of the R and S groups, while K^{RS} is a small "exchange correction." This result (discussed in greater detail in the appendix) provides a much firmer basis for a breakdown of the energy into bond and interbond terms—all with an essentially classical interpretation—but has yet to be developed along semiempirical lines (as it very well could be) for use in the situations studied during the early years of VB theory.

Concluding Remarks

In conclusion, we return to spin coupling and mention another area in which the "fictitious" spin couplings play a dominant role: This must serve to illustrate a vast untapped field of possible applications. Let us turn to the Heisenberg model and its variants, used in discussing ferromagnetic and antiferromagnetic interactions between molecules in a lattice. In such cases the spin Hamiltonian assumed is usually of the form

$$\mathsf{H}_S = C - \sum_{K<L} J_{KL} \mathbf{S}_K \cdot \mathbf{S}_L \tag{2.31}$$

which is analogous to the one due to Dirac and Van Vleck but in general may differ because $\mathbf{S}_K$ and $\mathbf{S}_L$ are *resultant* spins of systems K and L, and are not necessarily of magnitude $\frac{1}{2}$. The "exchange integral" J_{KL}, which is not always very clearly defined, is assumed to be capable of negative or positive values (positive for ferromagnetism), and the rest of the problem reduces to statistical mechanics. The quantum mechanical origin of the formula (2.31), in the context of ferromagnetism, is even now not completely clear, but we may establish the following result quite rigorously (including all nonorthogonality effects): For two systems A and B, with resultant spins $\mathbf{S}_A$ and $\mathbf{S}_B$, the energy, including interaction to first order, may be written as the expectation value of the spin Hamitonian

$$\mathsf{H}_S = a + b(\mathbf{S}_A \cdot \mathbf{S}_B) + c(\mathbf{S}_A \cdot \mathbf{S}_B)^2 + \cdots \qquad (2.32)$$

where b is of second degree in overlap integrals, c of the fourth degree, etc. Further reference to this result is made in the appendix, in view of many possible applications. Again we stress that no *physical* spin interactions are included at this stage. The energy

$$E = E_A + E_B + E_{\text{Coulombic}} + E_{\text{distortion}} + b\langle \mathbf{S}_A \cdot \mathbf{S}_B \rangle + \cdots \qquad (2.33)$$

has simply been recast in a form which indicates its dependence (as in the Heitler–London formula, to which it reduces for two *one*-electron systems) on the spin coupling. The coefficients depend on the charge densities, spin densities, etc., (to which we turn in a later lecture) in the separate systems and may be evaluated without difficulty. The Heitler–London formula, with nonorthogonality admitted, appears as a special case of this result.

We mention this formula not because of any implications for the theory of magnetism but because it has applications in many molecular situations where overlap is not too large. It may be applied widely to the bonding in transition metal complexes, to discussions of reactions between molecules in triplet states, and to a variety of spin-coupling effects of importance in ESR, luminescence, etc. The relevance of the formula depends simply on the fact that if one or both systems are in singlet states, the scalar product term vanishes, Coulomb repulsion (supplemented by strong distortion potentials) predominates, and there is no binding; but if both systems are in nonsinglet states, a strong attraction can occur when their resultant spins are anti-parallel coupled. The coordination of organic molecules to a transition metal is a case in point; when butadiene is bound to zero-valent platinum, for example, it seems that the ligand is in a state which is closer to an *excited triplet* state than to its normal ground state. The electronic situation may be visualized as the result of donation of an electron through a σ bond, followed by back donation through a π bond, the net result being a local promotion within each subsystem. This mechanism (reminiscent of that proposed by Chatt [*21*] and Dewar [*22*]) can be formulated in terms of a molecular orbital approximation [*23*] but receives a more general and elegant interpretation in terms of a formal spin coupling. There is in fact a considerable weight of experimental evidence (e.g., from geometry, force constants, etc.) to suggest that in such situations the condition of the ligand does strongly resemble that of the molecule in its first excited triplet state. Applications to the reactivity of molecules in triplet states, and in many other fields, are as yet unexplored.

It may be objected, at first sight, that such systems are of a degree of complexity far beyond anything envisaged in the early applications of spin-coupling methods, and that, if the VB method was unsatisfactory in dealing with simple molecules, we can hardly expect it to be more successful with platinum complexes. But this would be to miss the point: *we are no longer using the valence bond theory*, and are consequently no longer bound by its limitations and inconsistencies. Developments along these lines have in fact been made already in nonempirical *ab initio* calculations.

The valence bond theory in its original form seems to have finally left the stage: But it leaves behind a rich legacy of useful concepts and of theoretical methods, in all of which the coupling of spins plays a central, although somewhat formal, role. There are many aspects of the theory that merit further development; but these would lead us too far afield. It is time to leave this subject and turn to the "real" interactions that arise from the magnetic dipoles associated with spins. These are orders of magnitude smaller than the interactions involved in chemical bonding but, as we shall see, are amenable to direct observation by physical techniques and provide an abundance of information about the nature of the electron distribution.

REFERENCES

1. Pauling, L., and Wilson, E. B., "Introduction to Quantum Mechanics." McGraw-Hill, New York, 1935.
2. Van Vleck, J. H., *Rev. Mod. Phys.* **3**, 167 (1935).
3. Dirac, P. A. M., *Proc. Roy. Soc. (London) Ser. A* **123**, 714 (1929).
4. Sylvester, J. T., *Am. J. Math.* **1**, 64 (1878).
5. Simonetta, M., Gianinetti, E., and Vandoni, I., *J. Chem. Phys.* **48**, 1579 (1968).
6. McWeeny, R., *Symp. Faradaday Soc.* **2**, 7 (1968).
7. Kotani, M., Amemiya, A., Ishiguro, E., and Kimura, T., "Tables of Molecular Integrals." Maruzen, Tokyo, 1966.
8. Rumer, G., Teller, E., and Weyl, H., *Nachr. Akad. Wiss. Goettingen Math. Physik. Kl. IIa* **1932**, 449 (1932).
9. Penney, W. G., *Proc. Roy. Soc. (London) Ser. A* **158**, 306 (1937).
10. Moffitt, W., *Proc. Roy. Soc. (London) Ser. A* **199**, 487 (1949).
11. Barfield, M., *J. Chem. Phys.* **48**, 4458 (1968).
12. Karplus, M., and Anderson, D. H., *J. Chem. Phys.* **30**, 6 (1959).
13. Matsen, F. A., and Browne, J. C., *J. Phys. Chem.* **66**, 2332 (1963).
14. Slater, J. C., *J. Chem. Phys.* **19**, 220 (1951).
15. Löwdin, P.-O., *Phys. Rev.* **97**, 1474 (1955).
16a. McWeeny, R., *Proc. Roy. Soc. (London) Ser. A* **223**, 63 (1954).
16b. McWeeny, R., *Proc. Roy. Soc. (London) Ser. A* **223**, 306 (1954).
16c. McWeeny, R., *Proc. Roy. Soc. (London) Ser. A* **227**, 288 (1955).
17. Hurley, A. C., Lennard-Jones, J. E., and Pople, J. A., *Proc. Roy. Soc. (London) Ser. A* **220**, 446 (1953).
18. Parks, J. M., and Parr, R. G., *J. Chem. Phys.* **28**, 335 (1958).
19. McWeeny, R., *Proc. Roy. Soc. (London) Ser. A* **253**, 242 (1959).
20. Klessinger, M., and McWeeny, R., *J. Chem. Phys.* **42**, 3343 (1965).
21. Chatt, J., *J. Chem. Soc.* **1953**, 2939 (1953).
22. Dewar, M. J. S., *Bull. Soc. Chim. France* **18**, C71 (1951).
23. McWeeny, R., Mason, R., and Towl, A. C. D., *Discussions Faradaday Soc.* **1**, 20 (1969).

3

THE ORIGIN OF SPIN HAMILTONIANS:

A Simple Example

The Pauli theory of spin is "phenomenological." We have seen how it is possible to reproduce the simplest observed properties of spin by adding certain extra terms to the Hamiltonian operator and by including a "spin variable" in the wave function. But up to this point the origin of these extra terms remains obscure. Before going on to analyze some of the many observable effects to which they lead, it is interesting to sketch some of the subsequent developments that have led to a more detailed, but still incomplete, account of the "small terms" in the Hamiltonian. We shall then go on to show, within the context of a very simple example, how the effects of the small terms may be represented in an "effective" Hamiltonian (containing only *spin* operators) in much the same way that the bonding forces in the hydrogen molecule were represented formally (Lecture 2) in terms of an effective Hamiltonian containing a spin scalar product operator.

The Dirac Equation

The first step was taken by Dirac [*1*], who replaced the Hamiltonian function, which in the presence of an external magnetic field with vector potential **A** reads

$$H = [\mathbf{p} + (e/c)\,\mathbf{A}]^2/2m + V \tag{3.1}$$

by its relativistic counterpart

$$H = [m^2c^4 + (c\mathbf{p} + e\mathbf{A})^2]^{1/2} + V \tag{3.2}$$

and then tried to use the usual prescription, $H \to i\hbar\,\partial/\partial t$ and $p_x \to (\hbar/i)\,\partial/\partial x$, etc. to set up a wave equation. To avoid difficulties in interpreting the square root of the operator, and of nonlinearity of the resultant equation, Dirac wrote

$$(H - V)^2/c^2 - m^2c^4 - [\mathbf{p} + (e/c)\,\mathbf{A}]^2 = 0 \tag{3.3}$$

and assumed the possibility of an "operator factorization" such that

$$\left(\pi_0 + \sum_\mu \alpha_\mu \pi_\mu + \beta mc\right)\left(\pi_0 - \sum_\mu \alpha_\mu \pi_\mu - \beta mc\right) = 0 \tag{3.4}$$

should be equivalent to the preceding statement. The π's are

$$\pi_0 = (H - V)/c^2, \qquad \pi_\mu = p_\mu + (e/c)\,A_\mu \qquad (\mu = x, y, z) \tag{3.5}$$

and α_1, α_2, α_3, β are suitably defined operators. For equivalence of the two statements, the operators must have anticommutation properties reminiscent of those possessed by Pauli's operators. But with *four* instead of three, it is not possible to represent their effect in a two-dimensional spin space. In fact, a four-dimensional "spin space" is necessary, although use of the word "spin" does not yet imply any assumptions about its meaning, and instead of using a spin-orbital

$$\psi = \phi_\alpha \alpha + \phi_\beta \beta \tag{3.6}$$

to describe an electron, we must use a four-component function

$$\psi_D = \phi_1 e_1 + \phi_2 e_2 + \phi_3 e_3 + \phi_4 e_4 \tag{3.7}$$

When ψ_D is inserted as an operand the π's in (3.4) become operators working on the components $\phi_1, \ldots, \phi_4$ (functions of spatial variables) according to the usual rules, while $\alpha_1, \alpha_2, \alpha_3$, and β (whose properties are independent of spatial behavior, making no reference to the actual potentials, etc.) work on $e_1, \ldots, e_4$. Dirac used only the operator factor nearest the operand and thus obtained a wave equation (with a scalar product notation for the middle term)

$$(\pi_0 - \boldsymbol{\alpha} \cdot \boldsymbol{\pi} - \beta mc)\,\psi_D = 0$$

which may be written

$$(c\boldsymbol{\alpha} \cdot \boldsymbol{\pi} + V + \beta mc^2)\,\psi_D = i\hbar(\partial\psi_D/\partial t) \tag{3.8}$$

for comparison with the usual Schrödinger equation.

Because a two-component description of spin appears to be highly successful in a phenomenological approach, we suspect that the Dirac equation should be equivalent to a suitably constructed Pauli-type equation. There are in fact well-known methods for "projecting" a many-component equation into a representation of fewer dimensions, one of which we refer to presently, and it was not long before an equivalent Pauli equation was found:

$$\mathsf{H}\psi = i\hbar(\partial\psi/\partial t) \tag{3.9}$$

where

$$\mathsf{H} = \frac{1}{2m}\boldsymbol{\pi}^2 + V + \Big\{-\frac{\pi^4}{8m^3c^2} + \frac{e\hbar}{mc}\,\mathbf{S}\cdot\mathbf{B}$$
$$-\frac{e\hbar}{4m^2c^2}\Big(\mathbf{S}\cdot\boldsymbol{\pi}\times\mathbf{E} - \mathbf{S}\cdot\mathbf{E}\times\boldsymbol{\pi} - \frac{1}{2}\hbar\,\mathrm{div}\,\mathbf{E}\Big)\Big\} + \cdots \tag{3.10}$$

Here the scalar product of two vector operators (including $\mathbf{A}^2 = \mathbf{A}\cdot\mathbf{A}$) is interpreted as the sum

$$\mathbf{A}\cdot\mathbf{B} = \mathsf{A}_x\mathsf{B}_x + \mathsf{A}_y\mathsf{B}_y + \mathsf{A}_z\mathsf{B}_z$$

each vector operator comprising a set of three component operators (e.g., A_x, A_y, A_z). The cross product is a vector operator with components

$$(\mathbf{A} \times \mathbf{B})_x = \mathsf{A}_y\mathsf{B}_z - \mathsf{B}_y\mathsf{A}_z$$
$$\cdots$$

and the triple products may be interpreted accordingly in component form. The **S** operators are interpreted as Pauli operators and the wave function ψ as the *two*-component spin-orbital (3.6). Thus, one immediate success of the Dirac theory is its prediction of an energy term corresponding to a classical field-dipole interaction, with $\boldsymbol{\mu} = -(e\hbar/mc)\mathbf{S} = -2\beta\mathbf{S}$ which is the observed "anomalous" value[1] for electron spin. The other small terms (in the curly brackets) represent (i) a correction due to relativistic mass variation and (ii) the effects of spin-orbit interactions; we normally drop the former for low energy electrons, but include the latter terms, which are essentially those that would be expected from semiclassical considerations. As far as it goes (i.e., neglecting higher powers of small terms), this Hamiltonian seems to be fairly satisfactory in accounting for the observed facts.

The Breit Equation and Its Reduction

For a many-particle system, the situation is much more uncertain. Breit [*2*] set up a two-particle analog of the Dirac equation (with a wave function having 16 components instead of 4), and again it was possible to obtain a "Pauli approximation" by eliminating the small components. The resultant Hamiltonian contains the classically expected terms (e.g., a dipole-dipole interaction) plus some others, and appears to be fairly satis-

[1] A more advanced quantum electrodynamic calculation shows that the correct value is not 2 but 2.0023. In what follows, we usually use g for this "free-electron g value."

factory for two electrons. It may be applied to one electron and one nucleus by inserting an observed nuclear dipole ($\boldsymbol{\mu}_n = g_n \beta_n \mathbf{I}$) for one particle instead of $-g\beta\mathbf{S}$ and leads to the same spin-orbit coupling terms as the Dirac equation with the potential energy term taken to be that of a fixed nucleus.

For many electrons, and many nuclei, it is customary simply to add all the pairwise contributions. The more important terms, in the case where nuclear motion is neglected, may be collected to give a Pauli-type Hamiltonian $\mathsf{H} = \mathsf{H}_0 + \mathsf{H}'$ where H_0 is the spinless nonrelativistic Hamiltonian, while the "perturbation" is

$$\mathsf{H}' = \mathsf{H}_{\text{elec}} + \mathsf{H}_{\text{mag}} + \mathsf{H}_{\text{SL}} + \mathsf{H}_{\text{Z}} + \mathsf{H}_{\text{SS}} + \mathsf{H}_{\text{N}} \tag{3.11}$$

where H_{elec} arises from external electric fields, H_{mag} from external magnetic fields interacting with electronic orbital motion, H_{SL} from interaction between electron spin and orbital motion, H_{Z} from Zeeman interaction between electron spins and magnetic field, H_{SS} from electron spin-spin interactions, and H_{N} includes all "hyperfine" terms arising from nuclear spins. We have separated internal and external contributions to the fields, and now assume further that $\mathsf{H}_{\text{elec}} = 0$ (no *applied* electric field) and $\mathbf{A}(i) = \frac{1}{2}\mathbf{B} \times \mathbf{r}_i$ (uniform applied magnetic field). The detailed expressions for the various terms are listed below. (For a full discussion of the origin of these terms, the reader is referred to McWeeny and Sutcliffe [3].)

1. *Magnetic:*

$$\mathsf{H}_{\text{mag}} = \mathsf{H}'_{\text{mag}} + \mathsf{H}''_{\text{mag}} \tag{3.12}$$

where

$$\mathsf{H}'_{\text{mag}} = \beta \sum_i \mathbf{B} \cdot \mathsf{L}(i), \qquad \mathsf{H}''_{\text{mag}} = (e^2/8mc^2) \sum_i (\mathbf{B} \times \mathbf{r}_i)^2 \tag{3.13}$$

and the orbital angular momentum operator (about the origin of coordinates) is

$$\hbar \mathsf{L}(i) = \mathbf{r}_i \times \mathsf{p}(i) \tag{3.14}$$

The term H'_{mag} is responsible for the orbital paramagnetism of free atoms in states with nonzero angular momentum, but in second order also contributes to diamagnetism, reducing the main contribution which arises from $\mathsf{H}''_{\text{mag}}$: It also determines other effects as will be evident shortly.

2. *Spin-Orbit:*

$$\mathsf{H}_{\text{SL}} = g\beta^2 \Big[\sum_{n,i} \frac{Z_n \mathbf{S}(i) \cdot \mathbf{M}^n(i)}{r_{ni}^3} - \sum_{i,j}{}' \frac{2\mathbf{S}(i) \cdot \mathbf{M}^i(j) + \mathbf{S}(i) \cdot \mathbf{M}^j(i)}{r_{ij}^3}\Big] \tag{3.15}$$

where we have introduced a gauge-invariant angular momentum operator [cf. (3.14)]

$$\hbar \mathbf{M}^p(q) = \mathbf{r}_{pq} \times \boldsymbol{\pi}(q) \tag{3.16}$$

associated with the angular momentum of a particle at q about point p.

3. *Zeeman (Electronic):*

$$\mathsf{H}_{\text{Z}} = g\beta \sum_i \mathbf{B} \cdot \mathbf{S}(i) \tag{3.17}$$

This is the spin-field coupling, sometimes included in a "complete" Zeeman term of the form $\beta \sum_i \mathbf{B} \cdot (\mathbf{L}(i) + g\mathbf{S}(i))$. The orbital part arises, however, from H'_{mag} and is dealt with separately here.

4. *Electron Spin-Spin:*

$$\mathsf{H}_{\text{SS}} = -\frac{1}{2} g^2\beta^2 \sum_{i,j}{}' \Big[\frac{3(\mathbf{S}(i) \cdot \mathbf{r}_{ij})(\mathbf{S}(j) \cdot \mathbf{r}_{ij}) - r_{ij}^2 \mathbf{S}(i) \cdot \mathbf{S}(j)}{r_{ij}^5} + \frac{8\pi}{3} \delta(\mathbf{r}_{ij})\, \mathbf{S}(i) \cdot \mathbf{S}(j)\Big] \tag{3.18}$$

This is the dipole-dipole coupling between electronic dipoles, supplemented by a "contact term" which in most cases is very small and produces no readily observable effects.

5. *Hyperfine:*

$$\mathsf{H}_N = -\sum_n g_n\beta_n \mathbf{B}\cdot \mathbf{I}(n) + 2\beta \sum_{n,i} g_n\beta_n r_{ni}^{-3}\mathbf{I}(n)\cdot \mathbf{M}^n(i)$$
$$+ g\beta \sum_{n,i} g_n\beta_n \Big\{ r_{ni}^{-5}[3(\mathbf{S}(i)\cdot \mathbf{r}_{ni})(\mathbf{I}(n)\cdot \mathbf{r}_{ni}) - r_{ni}^2 \mathbf{I}(n)\cdot \mathbf{S}(i)]$$
$$+ \frac{8\pi}{3}\,\delta(\mathbf{r}_{ni})\,\mathbf{I}(n)\cdot \mathbf{S}(i)\Big\} + \mathsf{H}_{NN} \qquad (3.19)$$

This contains, respectively, a *nuclear* Zeeman term, a term representing the nuclear magnetic moment acting on the moving electrons, and a dipole-dipole interaction with a contact term which in this case leads to important observable effects in NMR and ESR. Finally, H_{NN} includes a direct nuclear-nuclear, dipole-plus-contact coupling.

Other terms may be added (e.g., to allow for nuclear quadrupole moments, nuclear motion, etc.) but will not be considered here. The first four terms give all the specifically electronic effects, while H_N is responsible for all further nuclear hyperfine effects, including the nuclear modification of ESR signals and, of course, the whole of NMR spectroscopy.

The Spin Hamiltonian: Free Atom

Now that we know (or think we know, with adequate accuracy) the small terms in the Hamiltonian, we may pick up the threads where we left them in Lecture 1. What observable effects do the spins produce? We are armed with two useful ideas, which we shall develop at length, namely (i) that the spin interactions will split levels that would otherwise be degenerate, giving them a fine structure, and (ii) that it is not always necessary to solve a complete many-electron Schrödinger equation to "fit" the observed fine structure, since when the splitting is linked with spin (even in a purely formal way), we may be able to set up an *effective Hamiltonian*, containing only spin operators and numerical parameters, whose eigenvalues can reproduce the observed levels. In fact the idea of fitting the observed levels in magnetic resonance experiments by means of a "spin Hamiltonian" came long before its detailed interpretation in terms of molecular wave functions. The spin Hamiltonian has been called the "last outpost" of the experimentalist [*4*], the point at which all the observations have been formally interpreted and the results put into one small package, ready for the theoretician to start relating coupling constants and other parameters to the actual form of the electron distribution.

We shall introduce the theme of the remaining lectures by means of one simple and somewhat hypothetical example, going back to the ^{2}P state of the sodium atom which we considered in Lecture 1, and applying a strong crystal field to "quench" the orbital angular momentum and break its coupling to the spin. The resultant splitting of the energy levels will exhibit the main features of that encountered in dealing with transition metals in complex ions, as revealed experimentally in their ESR spectra.

First, we recall that for the free atom the orbital and spin angular momenta ($L = 1$, $S = \frac{1}{2}$) are coupled to give a resultant with quantum number $J = \frac{1}{2}$ or $\frac{3}{2}$: the states $^2P_{1/2}$ and $^2P_{3/2}$ are resolved into $2J + 1$ components (i.e., 2 and 4, respectively) on applying a magnetic field **B**, and the effective magnetic moment (along the field direction, z) associated with a state of angular momentum J is $g\beta \mathrm{J}_z$, where the Landé factor[2] g takes account of the coupling and is

$$g = 1 + \frac{J(J+1) + S(S+1) - L(L+1)}{2J(J+1)} \tag{3.20}$$

[2] Note that this *effective* g value, including an orbital contribution, may take a value very different from that for a free electron (although we have used the same symbol).

Atoms in states with $J = \frac{1}{2}$ and $\frac{3}{2}$ simply behave like magnets with $\mu_z = g\beta \mathsf{J}_z$ to give the splitting of levels shown in Fig. 1.4.

The complexities of a direct calculation, involving both spin-orbit coupling and the interaction of the magnetic moment with the external field, may now be avoided by saying that the energies of the Zeeman levels from any one "parent" (energy E_J) are the expectation values of an effective Hamiltonian

$$\mathsf{H}_{\text{eff}} = E_J + g\beta B \mathsf{J}_z$$

where g is a suitably chosen parameter ($g = \frac{2}{3}$ or $\frac{4}{3}$ for $J = \frac{1}{2}$ or $\frac{3}{2}$). Since J_z has expectation values $M_J = J, J - 1, ..., -J$, the energy levels follow at once.

We could even go further, and pretend there was no orbital angular momentum at all, by regarding J_z as a *fictitious* spin operator, for an imaginary system in which the total spin had the value $S = J$ but a magnetic moment $g\beta\mathsf{S}_z$ instead of $2\beta\mathsf{S}_z$. In this case each level has its own spin Hamiltonian

$$\mathsf{H}_{\text{s}}^{(1/2)} = E_{1/2} + \tfrac{2}{3}\beta B \mathsf{S}_z', \qquad \mathsf{H}_{\text{s}}^{(3/2)} = E_{3/2} + \tfrac{4}{3}\beta B \mathsf{S}_z''$$

so that putting in the expectation values of S_z' in spin eigenstates Θ_M' (with $M = \pm\frac{1}{2}$ corresponding to fictitious spin $S' = \frac{1}{2}$) reproduces the Zeeman energies of the doublet, while those of the quartet are similarly reproduced as expectation values in fictitious spin states Θ_M'' (with $M = \pm\frac{1}{2}, \pm\frac{3}{2}$ corresponding to fictitious spin $S'' = \frac{3}{2}$). The effects of orbital motion in the states with $J = \frac{1}{2}, \frac{3}{2}$ are thus merely reflected in the different parameter values in the spin-Hamiltonian description of the effect of an applied magnetic field.

Effect of Crystal Fields

We now put the system in a strong tetragonal field (e.g., by bringing up four charged ligands as in Fig. 3.1a) and try to formulate a similar description of the resultant Zeeman levels. The orbital motion may be "impeded" by a tetragonal field, and the corresponding angular momentum quenched; and if only *spin* angular momentum remains, we might expect that the spin-field coupling could again be described by a term of the form $g\beta BS_z$ as in (1.17). But would it be correct to use the spin-only value of $g = 2$? And is there any real basis for considering a spin-only Hamiltonian instead of the full one in this more general case?

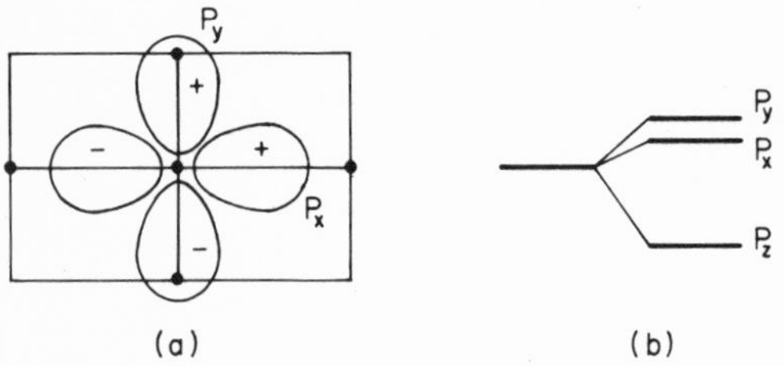

FIG. 3.1. Crystal field splitting of a p level. (a) Charged ligands (assumed negative) approaching along x and y axes, with (real) p_x, p_y orbitals. (b) Splitting of initially degenerate level.

The first step is to get appropriate orbitals to describe the single valence electron *in the tetragonal field.* For the free atom, we would most naturally use p_{+1}, p_0 and p_{-1} (with $M_L = +1$, 0, -1); but if we put on the field, we find these are no longer appropriate; by setting up a secular problem (or by symmetry considerations), we find they are mixed to give p_0, $(p_{+1} + p_{-1})/\sqrt{2}$, and $(p_{+1} - p_{-1})/\sqrt{2}$. In more familiar form these are p_z, p_x, and p_y, the *real* dumbell-shaped orbitals. It is easy to verify that the orbital angular momentum has been quenched by calculating the expectation value of each component—each one vanishes. The energy expectation values may be calculated by averaging the imposed potential felt by an electron in each of the three states, i.e., distributed according to p_z^2, p_x^2, or p_y^2. Thus, if the ligands put negative charge near the metal, the energies of the p_x and p_y states will be raised above that of the p_z. The threefold orbital degeneracy of the free atom is thus resolved completely and the lowest state will be a doublet when spin is included. The situation is now as depicted in Fig. 3.1b and we are ready to add the spin terms in the Hamiltonian.

In spin resonance experiments we resolve the ground state doublet into its Zeeman components and measure the splitting. We would like to concentrate on these two functions alone since we know that under a perturbation these degenerate functions will be mixed and the resolved energy levels will be obtained by solving a 2×2 secular problem, but we know that there will be some mixing of *all* the functions that might be used in expanding the wave function. We therefore look at the situation in a more general way.

The Effective Hamiltonian

We suppose the functions before applying the spin and magnetic field perturbations are $\Phi_1, \Phi_2, \ldots, \Phi_\kappa, \ldots$, and that these are approximate eigenfunctions of the corresponding Hamiltonian H_0. If the wave function is expanded in the form

$$\Phi = \sum_\kappa c_\kappa \Phi_\kappa \tag{3.21}$$

the perturbed states, for $\mathsf{H}_0 \rightarrow \mathsf{H} = \mathsf{H}_0 + \mathsf{H}'$, are given by solving the secular equations

$$\begin{aligned} H_{11}c_1 + H_{12}c_2 + \cdots &= Ec_1 \\ H_{21}c_1 + H_{22}c_2 + \cdots &= Ec_2 \\ &\cdots \end{aligned}$$

The general matrix element is

$$H_{\kappa'\kappa} = \langle \Phi_{\kappa'} \mid \mathsf{H} \mid \Phi_\kappa \rangle \tag{3.22}$$

and the secular equations may be written in matrix form as

$$\mathbf{Hc} = E\mathbf{c} \tag{3.23}$$

In our example we want to concentrate on Φ_1 and Φ_2 (which are, here, single spin-orbitals, $p_z\alpha$ and $p_z\beta$), and expect that Φ_3, Φ_4, Φ_5, Φ_6 will be less important; we therefore distinguish the two sets of functions as the "A group" and the "B group" and partition the matrix **H** and column **c** accordingly. It then follows readily that the first two solutions of the full secular equations will be identical with those of the 2×2 secular problem

$$\mathbf{H}_{\text{eff}}\mathbf{a} = E\mathbf{a} \tag{3.24}$$

in which $\mathbf{H}_{\text{eff}}$ is an *effective* Hamiltonian matrix. This result follows most elegantly [5] on *partitioning* the matrix equation (3.23) in the form

$$\begin{pmatrix} \mathbf{H}^{AA} & \mathbf{H}^{AB} \\ \mathbf{H}^{BA} & \mathbf{H}^{BB} \end{pmatrix} \begin{pmatrix} \mathbf{a} \\ \mathbf{b} \end{pmatrix} = E \begin{pmatrix} \mathbf{a} \\ \mathbf{b} \end{pmatrix}$$

It is then equivalent to two equations

$$\mathbf{H}^{AA}\mathbf{a} + \mathbf{H}^{AB}\mathbf{b} = E\mathbf{a}$$

$$\mathbf{H}^{BA}\mathbf{a} + \mathbf{H}^{BB}\mathbf{b} = E\mathbf{b}$$

and, for $E \simeq E_a$ (the energy of the degenerate A-group states), the second equation allows us to estimate the column **b**, and then to obtain the "correction term" $\mathbf{H}^{AB}\mathbf{b}$ in the first. The revised form of the first equation is essentially (3.24) in which the matrix $\mathbf{H}_{\text{eff}}$ takes account of intergroup mixing to a well-defined order (which may be pushed higher and higher by iteration). The result is quite simple: If we denote the A- and B-group functions by $\{\Phi_{\kappa_a}\}$ and $\{\Phi_{\kappa_b}\}$, respectively, and write the required elements of $\mathbf{H}_{\text{eff}}$ formally as $\langle\kappa_a' \mid \mathsf{H}_{\text{eff}} \mid \kappa_a\rangle$, then we find

$$\langle\kappa_a' \mid \mathsf{H}_{\text{eff}} \mid \kappa_a\rangle = \delta_{\kappa_a'\kappa_a} E_a + \langle\kappa_a' \mid \mathsf{H}' \mid \kappa_a\rangle$$

$$+ \sum_{\kappa_b} \frac{\langle\kappa_a' \mid \mathsf{H}' \mid \kappa_b\rangle \langle\kappa_b \mid \mathsf{H}' \mid \kappa_a\rangle}{E_a - E_{\kappa_b}} + \cdots \qquad (3.25)$$

We have assumed that the A-group (states $\Phi_{\kappa_a}, \Phi_{\kappa_a'}$) are degenerate with energy E_a, and in this case the first term on the right may be dropped if we measure all energies relative to E_a. We are usually content to drop the higher-order terms (not shown), although they can be added if necessary. This result tells us how to achieve our object of concentrating only on one multiplet level of interest, absorbing the wider effects of the perturbation by using a suitably modified Hamiltonian. It does not yet tell us how to get a pure *spin* Hamiltonian, but it is a step in the right direction, and we apply it to the system under consideration.

Detailed Solution

Let us now use $\mid M_L M_S\rangle$ to denote states of the electron in the free atom (so that, e.g., $\text{p}_{+1}\alpha = \mid 1, \tfrac{1}{2}\rangle$) and work out the effective Hamiltonian. The unperturbed states for the "complex" are represented by $\text{p}_0\alpha$, $\text{p}_0\beta$ (A group) and $\text{p}_x\alpha$, $\text{p}_x\beta$, $\text{p}_y\alpha$, $\text{p}_y\beta$ (B group). We also consider, for simplicity, a square planar situation: In this case p_x, p_y are degenerate and we may as well stick to the alternative linear combinations $\text{p}_{\pm 1}\alpha$, $\text{p}_{\pm 1}\beta$ which are slightly more convenient to handle. We therefore consider, as unperturbed functions,

Φ_1	Φ_2	Φ_3	Φ_4	Φ_5	Φ_6
$\mid 0, \tfrac{1}{2}\rangle$	$\mid 0, -\tfrac{1}{2}\rangle$	$\mid 1, \tfrac{1}{2}\rangle$	$\mid 1, -\tfrac{1}{2}\rangle$	$\mid -1, \tfrac{1}{2}\rangle$	$\mid -1, -\tfrac{1}{2}\rangle$

the first set being the Φ_{κ_a}, the second set being the Φ_{κ_b}. To simplify matters a little, we use an approximate form of the spin-orbit operator (3.15) (which turns out to be accurately valid for a strictly central field), namely $H_{SL} = \lambda \mathbf{L} \cdot \mathbf{S}$, and we also include (as we obviously must) the direct interaction between the field and the spin and orbital angular momenta [H_Z and H'_{mag} in (3.17) and (3.13)]. The perturbation to be considered is thus

$$H' = \lambda \mathbf{L} \cdot \mathbf{S} + \beta \mathbf{B} \cdot (\mathbf{L} + 2\mathbf{S}) = \lambda H_1 + \beta H_2 \tag{3.26}$$

which we assume contains the main effects of electron spin and applied field. Now each scalar product may be rewritten in terms of step-up and step-down operators:

$$\mathbf{L} \cdot \mathbf{S} = \tfrac{1}{2}(L^+S^- + L^-S^+) + L_zS_z$$

$$\mathbf{B} \cdot \mathbf{L} = \tfrac{1}{2}(B^+L^- + B^-L^+) + B_zL_z$$

$$\mathbf{B} \cdot \mathbf{S} = \tfrac{1}{2}(B^+S^- + B^-S^+) + B_zS_z$$

(where $B^{\pm} = B_x \pm iB_y$) and it is therefore easy to evaluate the matrix elements in the formula (3.25) for $\langle \kappa_a' \mid H' \mid \kappa_a \rangle$. In general, using the M_LM_S labels, these may be denoted by $\langle M_L', M_S' \mid H' \mid M_L, M_S \rangle$, as we use $\mid M_L, M_S \rangle$ for the states.

First we obtain the A-group matrix elements, with $M_L = M_L' = 0$. Thus,

$$\langle 0, \tfrac{1}{2} \mid H' \mid 0, \tfrac{1}{2} \rangle = \beta B_z, \qquad \langle 0, -\tfrac{1}{2} \mid H' \mid 0, -\tfrac{1}{2} \rangle = -\beta B_z \tag{3.27a}$$

because the step operators spoil the matching of the quantum numbers on the two sides of H', and hence give zero by orthogonality, while the L_z term also gives zero for states with $M_L = 0$. The off-diagonal elements $\langle 0, \tfrac{1}{2} \mid H' \mid 0, -\tfrac{1}{2} \rangle$ and $\langle 0, -\tfrac{1}{2} \mid H' \mid 0, \tfrac{1}{2} \rangle$ contain nonzero terms from operators that can step the spin up or down by one unit without changing the orbital quantum number and clearly yield βB^- and βB^+, respectively. We find, in fact,

$$\langle 0, \tfrac{1}{2} \mid \mathsf{H}' \mid 0, -\tfrac{1}{2}\rangle = \beta B^-, \qquad \langle 0, -\tfrac{1}{2} \mid \mathsf{H}' \mid 0, \tfrac{1}{2}\rangle = \beta B^+ \quad (3.27b)$$

Next we note that the second-order sum in (3.25), with $\mathsf{H}' = \lambda \mathsf{H}_1 + \beta \mathsf{H}_2$, will give terms quadratic in λ (and hence the spins), quadratic in β (and hence the field components), and bilinear in λ and β. It is the latter terms that give energy changes proportional to the field, as observed in the normal Zeeman effect, the other terms leading to a spin-spin coupling and to diamagnetism, which do not concern us at the moment. We therefore keep only the following parts of the second-order sum:

$$\lambda\beta \sum_{M_L, M_S''} (E_0 - E_{M_L})^{-1} [\langle 0, M_S' \mid \mathsf{H}_1 \mid M_L, M_S''\rangle \langle M_L, M_S'' \mid \mathsf{H}_2 \mid 0, M_S\rangle$$
$$+ \langle 0, M_S' \mid \mathsf{H}_2 \mid M_L, M_S''\rangle \langle M_L, M_S'' \mid \mathsf{H}_1 \mid 0, M_S\rangle]$$

where M_L and M_S'' refer to the intermediate states with $M_L = \pm 1$, $M_S'' = \pm\frac{1}{2}$.

Let us take as a first example the contribution to $\langle 0, M_S' \mid \mathsf{H}_{\text{eff}} \mid 0, M_S\rangle$ arising when $M_S = M_S' = \frac{1}{2}$. The only non-zero H_1 terms in the square bracket are then

$$\langle 0, \tfrac{1}{2} \mid \mathsf{H}_1 \mid 1, -\tfrac{1}{2}\rangle \qquad \text{or} \qquad \langle 1, -\tfrac{1}{2} \mid \mathsf{H}_1 \mid 0, \tfrac{1}{2}\rangle$$

where the intermediate state $\mid M_L, M_S\rangle$ can only differ from $\mid 0, \frac{1}{2}\rangle$ by simultaneous step *up* of one unit in M_L, and step *down* of one unit in M_S. The actual values of the elements are easily obtained from the standard formula (for orbital states $\mid M_L\rangle$, with a similar result for spin states $\mid M_S\rangle$)

$$\mathsf{L}^{\pm} \mid M_L\rangle = [(L \mp M)(L \pm M + 1)]^{1/2} \mid M_L \pm 1\rangle$$

On the other hand, the matrix element $\langle 1, -\frac{1}{2} \mid \mathsf{H}_2 \mid 0, \frac{1}{2}\rangle$ vanishes because H_2 does not contain operators capable of changing M_L and M_S simultaneously. The sum thus contains no nonzero terms. When we turn to the off-diagonal elements, however, we find a contribution to $\langle 0, \frac{1}{2} \mid \mathsf{H}_{\text{eff}} \mid 0, -\frac{1}{2}\rangle$ of the form

$$\lambda\beta(E_0 - E_1)^{-1} [\langle 0, \tfrac{1}{2} \mid \mathsf{H}_1 \mid 1, -\tfrac{1}{2}\rangle \langle 1, -\tfrac{1}{2} \mid \mathsf{H}_2 \mid 0, -\tfrac{1}{2}\rangle + \langle 0, \tfrac{1}{2} \mid \mathsf{H}_2 \mid -1, \tfrac{1}{2}\rangle \langle -1, \tfrac{1}{2} \mid \mathsf{H}_1 \mid 0, -\tfrac{1}{2}\rangle]$$

which easily reduces, denoting the "crystal field" splitting by $(E_1 - E_0) = \Delta$, to $-2\lambda\beta \mathrm{B}^-/\Delta$.

On putting the contributions together, we find, for the terms linear in the applied field,

$$\begin{aligned} \langle 0, \tfrac{1}{2} \mid \mathsf{H}_{\text{eff}} \mid 0, \tfrac{1}{2}\rangle &= \beta B_z \\ \langle 0, \tfrac{1}{2} \mid \mathsf{H}_{\text{eff}} \mid 0, -\tfrac{1}{2}\rangle &= \beta B^- - 2\lambda\beta B^-/\Delta \\ \langle 0, -\tfrac{1}{2} \mid \mathsf{H}_{\text{eff}} \mid 0, \tfrac{1}{2}\rangle &= \beta B^+ - 2\lambda\beta B^+/\Delta \\ \langle 0, -\tfrac{1}{2} \mid \mathsf{H}_{\text{eff}} \mid 0, -\tfrac{1}{2}\rangle &= -\beta B_z \end{aligned} \tag{3.28}$$

The splitting of the degenerate doublet by the combined effect of spin-orbit coupling and applied field is thus determined by solving the reduced secular problem (3.24), which refers explicitly to the A group alone and takes the form

$$\begin{pmatrix} \beta B_z & (1 - 2\lambda/\Delta)\,\beta B^- \\ (1 - 2\lambda/\Delta)\,\beta B^+ & -\beta B_z \end{pmatrix} \begin{pmatrix} a_1 \\ a_2 \end{pmatrix} = E \begin{pmatrix} a_1 \\ a_2 \end{pmatrix} \tag{3.29}$$

If the field is along the z axis, $B^+ = B^- = 0$ and we obtain at once $E = \pm\beta B_z$. The splitting is then $\Delta E = g\beta B_z$ with $g = 2$, just as if we had a free electron with no orbital angular momentum at all. But if the field is along, say, the x axis, so that $B_z = 0$ and $B^- = B^+ = B_x$, the resultant secular equations give a consistency condition (secular determinant)

$$\begin{vmatrix} -E & X \\ X & -E \end{vmatrix} = 0$$

with $X = (1 - 2\lambda/\Delta)\beta B_x$. The levels are then

$$E = \pm(1 - 2\lambda/\Delta)\,\beta B_x$$

and the electron is consequently behaving as if its g value were

$$g = 2(1 - 2\lambda/\Delta)$$

instead of 2. What is happening is that, even though a crystal field may quench the angular momentum, the applied magnetic field can slightly upset the balancing of the p_{+1} and p_{-1} in the real combinations p_x and p_y. If the weights change even slightly we find a small field-induced angular momentum which, through the spin-orbit coupling term, may reduce or augment the value due to spin only, depending on the sign of the coupling parameter λ.

The Spin Hamiltonian

The final step in this discussion, and the one on which the whole idea of the spin Hamiltonian depends, consists of showing that a "model" system can be set up in such a way as to reproduce exactly the behavior of the real system. The model system should comprise only a spin (with spin operators S_x, S_y, S_z) interacting with the applied field **B**, and the Hamiltonian of this fictitious system should include nothing else but numerical parameters. To see how this goal can be achieved, we need only look at the matrix elements of the effective Hamiltonian [appearing on the left in (3.29)] and compare them with those of the spin operators S_z, S^+, and S^- (together with the unit operator I) taken between spin states α and β. The latter yield matrices

$$\mathbf{S}_z = \begin{pmatrix} \frac{1}{2} & 0 \\ 0 & -\frac{1}{2} \end{pmatrix} \qquad \mathbf{S}^+ = \begin{pmatrix} 0 & 1 \\ 0 & 0 \end{pmatrix} \qquad \mathbf{S}^- = \begin{pmatrix} 0 & 0 \\ 1 & 0 \end{pmatrix} \qquad \mathbf{1} = \begin{pmatrix} 1 & 0 \\ 0 & 1 \end{pmatrix}$$

respectively, and it is clear that any 2×2 matrix may be written as a linear combination of these four matrices. In particular, the matrix $\mathbf{H}_{\text{eff}}$ on the left-hand side of (3.29) may be written

$$\mathbf{H}_{\text{eff}} = 2\beta B_z \mathbf{S}_z + (1 - 2\lambda/\Delta)\,\beta B^- \mathbf{S}^+ + (1 - 2\lambda/\Delta)\,\beta B^+ \mathbf{S}^-$$

The rest of the argument is obvious and may be summarized as follows: If we set up a *spin Hamiltonian*

$$\mathsf{H}_{\mathsf{S}} = 2\beta B_z \mathsf{S}_z + (1 - 2\lambda/\Delta)\,\beta B^- \mathsf{S}^+ + (1 - 2\lambda/\Delta)\,\beta B^+ \mathsf{S}^- \qquad (3.30)$$

this will yield a secular problem, in a basis of spin states, identical with that which we reached in (3.29) by a more complete calculation based on the actual Hamiltonian. This is the crux of the matter, and the possibility of "simulating" the behavior of an actual system by means of a fictitious spin system is the key idea in fitting experimental results to a phenomenological spin Hamiltonian.

In Cartesian form the spin Hamiltonian (3.30) may be written

$$\mathsf{H}_{\mathsf{S}} = \beta \sum_{\lambda,\mu} g_{\lambda\mu} B_\lambda \mathsf{S}_\mu \qquad (\lambda, \mu = x, y, z) \qquad (3.31)$$

where the $g_{\lambda\mu}$ comprise the *g tensor* of the system. In the present case

$$g_{zz} = 2, \qquad g_{xx} = g_{yy} = 2(1 - 2\lambda/\Delta)$$

Vanishing of the other components means that the axes chosen are the principal axes of the tensor, which normally coincide with symmetry axes of the system.

Experimental Implications

Clearly, the experimental determination of a *g* tensor, from observation of the dependence of the Zeeman splittings on field direction (using ESR techniques), can give valuable information about the orientation of molecules or complex ions in a crystal, provided that free spins are available. This is the case, for example, with transition metals containing incompletely filled d orbitals, and the analysis we have used may be taken over with very little change for a system such as V^{4+} with one d electron. One well-known and significant example from the earlier work in this field was the determination of the orientation of the heme groups in crystallized hemoglobin derivatives from the g tensor of the ferric ion, and of similar groups (containing magnesium instead of iron) in chlorophyll [*6*, *7*]. In each case the ion lies at the center of a phthalocyanine structure which forms only a part of a very large structure (molecular weight 20,000!), and yet we are able to "see" the spin distribution associated with a few valence electrons on one atom in the molecule.

We now begin to see the immense potentialities of using spins as indicators of the electron distribution in molecules; they can provide not only structural information (which can be obtained even without a detailed appreciation of the underlying quantum mechanics), but also evidence concerning the nature of chemical bonding, the effect of ligands on oxidation numbers, etc., and the validity of different descriptions proposed by the theoretical chemist. This evidence requires a careful interpretation of the connection between the spin Hamiltonian parameters obtained by the experimentalist and the electronic wave function proposed by the theoretician—issues to which we turn in the next lecture.

REFERENCES

1. Dirac, P. A. M., *Proc. Roy. Soc. (London) Ser. A* **117**, 610 (1928).
2. Breit, G., *Phys. Rev.* **35**, 1447 (1930).
3. McWeeny, R., and Sutcliffe, B. T., "Methods of Molecular Quantum Mechanics." Academic Press, New York, 1969.
4. Griffith, J. S., "The Theory of Transition-Metal Ions." Cambridge Univ. Press, London and New York, 1961.
5. Löwdin, P.-O., *J. Chem. Phys.* **19**, 1396 (1951).
6. Ingram, D. J. E., and Bennett, J. E., *J. Chem. Phys.* **22**, 1136 (1954).
7. Bennett, J. E., and Ingram, D. J. E., *Nature* **175**, 130 (1955).

4

SPIN DENSITY, SPIN POPULATIONS, AND SPIN CORRELATION

We have seen that the small terms in the Hamiltonian may be treated by perturbation theory, that the theory refers formally only to a degenerate group of states—whose resolution into field- and spin-dependent components is what really concerns us—and that there is some basis for discussing the problem in terms of an effective Hamiltonian containing *spin operators* only using a basis consisting of *spin functions* only. There is thus some theoretical justification for the use of a "phenomenological" spin Hamiltonian in fitting observed results. We now wish to know whether such a procedure is valid generally and, even more important, what information we can extract from a list of experimentally determined parameters such as g values and coupling constants. What can these numbers tell us about the shapes and orientations of molecules, the electronic structure of the bonds, and the distribution of any "unpaired" spins? Here we need new *concepts* to bridge the gap between basic theory, which gives us in the first place wave functions, and the interpretation of experiments. Crystallography is well served in this respect; the crystallographer is not worried about the details of the electronic wave function and is quite happy to work in terms of a derived concept—the electron density. In the interpretation of spin-Hamiltonian parameters it is important to look for analogous functions, not merely because they provide a convenient physical picture of the electron distribution but really because there is no other sensible way of proceeding. Molecular wave functions become more and more elaborate as computing facilities improve, and a function good enough to predict spin properties with high accuracy might contain hundreds of spin-orbital determinants, each orbital being an optimized linear combination of atomic orbitals or other suitable "basis functions." We do not always wish to mount a full-scale calculation in order to interpret a spin-spin coupling constant (and the calculation would probably yield a result not very close to the experimental value, the wave function not being good enough). More probably, we want to get some appreciation of how a particular coupling constant is connected with the general features of the electron distribution—not with the wave function itself, which contains a vast amount of superfluous information, but with simpler "density functions" that will tell us where the electron density is high, or where a free spin is most likely to be found, or whether spins of electrons at A and B are more likely to be coupled parallel or antiparallel. We are faced, therefore, with the general problem of obtaining information about the electron distribution and how it determines molecular properties, and we must consider this at some length before we can begin to discuss in detail the physical meaning of spin-Hamiltonian parameters.

Electron Density Functions

We are usually concerned with orbital wave functions. These are constructed by allocating electrons to orbitals A, B,..., R,..., with spin factors α or β, and forming antisymmetrical functions from them so as to satisfy the Pauli principle. Each orbital with its spin factor is a spin-orbital (e.g., $A\alpha$, $A\beta$), and any specification of spin-orbitals for all the electrons concerned is a *spin-orbital configuration*, with which we can associate a single antisymmetric function—a Slater determinant. Many such determinants may have to be mixed, i.e., put together with suitable coefficients, to get a good wave function. In molecular orbital theory the orbitals are nonlocalized (MO's), and one configuration of doubly occupied orbitals $A\alpha$, $A\beta$, $B\alpha$, $B\beta$,..., may give a fairly good wave function if the forms of the MO's are carefully chosen. In valence bond (VB) theory, on the other hand, the orbitals are localized AO's, and many spin-orbital configurations must be combined to give a wave function of comparable accuracy. At the moment, however, we are concerned with the general properties of any kind of wave function, however complicated.

First we recall the notation. An orbital is a function of position in space (position vector $\mathbf{r}$) and its squared value at $\mathbf{r}$, $|A(\mathbf{r})|^2$ $[=A^*(\mathbf{r})\,A(\mathbf{r})]$, indicates the probability of finding the electron there. The spin factor α or β, is written formally as a function of spin component s $(=S_z)$ along an axis of quantization (conventionally taken as z axis), the interpretation of the functional notation being the one we used in Lecture 2 (p. 65). Thus, $\alpha(s)$ vanishes unless $s = \frac{1}{2}$, and this means that the spin component has zero probability of being different from $\frac{1}{2}$ (i.e., half an atomic unit) in state α, while $\beta(s)$ vanishes unless $s = -\frac{1}{2}$, corresponding to certainty of finding spin $-\frac{1}{2}$ in state β.

The state of a single electron is then described by a spin orbital. We denote space and spin variables $(\mathbf{r}, s)$ collectively by $\mathbf{x}$, and the spin-orbital $\psi_A(\mathbf{x}) = A(\mathbf{r})\,\alpha(s)$ then describes an electron in orbital A "with spin $+\frac{1}{2}$." The statistical interpretation is simple. The probability of finding the electron in element $d\mathbf{r}$ and with spin between s and $s + ds$ is determined by a *density function* $\rho(\mathbf{x})$ such that

$$\text{Probability of finding electron in } d\mathbf{x} = \rho(\mathbf{x})\,d\mathbf{x} = |\psi_A(\mathbf{x})|^2\,d\mathbf{x}$$
$$= |A(\mathbf{r})|^2\,|\alpha(s)|^2\,d\mathbf{r}\,ds$$

and is zero unless s is in the vicinity of $+\frac{1}{2}$, since this is a "plus-spin" electron. If we are not interested in spin but only in where the electron is, we can sum over all spin possibilities and obtain, for the probability of the electron being found in $d\mathbf{r}$,

$$P(\mathbf{r})\, d\mathbf{r} = \left(\int \rho(\mathbf{x})\, ds\right) d\mathbf{r} = |\, A(\mathbf{r})\,|^2\, d\mathbf{r}$$

just as if the electron had no spin and was simply put into orbital A. The function $\rho(\mathbf{x})$ is a probability density including the spin description, while

$$P(\mathbf{r}) = \int \rho(\mathbf{x})\, ds$$

is a probability density without reference to spin and is obtained by summing (integrating) over spin. These are the prototypes of functions that can be defined for a many-electron system. P is particularly useful because, for many purposes, it is possible to regard the electron as actually "smeared out" with density P.

We now generalize to the case of many electrons,[1] using $\mathbf{x}_i$ for the space spin coordinates of electron i and $\mathbf{r}_i$, s_i for space and spin separately. The wave function is $\Psi(\mathbf{x}_1, \mathbf{x}_2, \ldots, \mathbf{x}_N)$ and has the interpretation

$$\Psi(\mathbf{x}_1, \mathbf{x}_2, \ldots, \mathbf{x}_N)\, \Psi^*(\mathbf{x}_1, \mathbf{x}_2, \ldots, \mathbf{x}_N)\, d\mathbf{x}_1\, d\mathbf{x}_2 \cdots d\mathbf{x}_N$$
$$= \text{Probability of electron 1 in } d\mathbf{x}_1,\ 2 \text{ simultaneously in } d\mathbf{x}_2,\ \cdots$$

The probability of 1 in $d\mathbf{x}_1$ and other electrons *anywhere* is thus

$$d\mathbf{x}_1 \int \Psi(\mathbf{x}_1, \mathbf{x}_2, \ldots, \mathbf{x}_N)\, \Psi^*(\mathbf{x}_1, \mathbf{x}_2, \ldots, \mathbf{x}_N)\, d\mathbf{x}_2 \cdots d\mathbf{x}_N$$

and the probability of finding *any* of the N electrons in $d\mathbf{x}_1$ is N

[1] The exposition and notation follow closely that of McWeeny [*1a–c*].

times this. We write this probability as $\rho_1(\mathbf{x}_1)\, d\mathbf{x}_1$, where the density function $\rho_1(\mathbf{x}_1)$ is

$$\rho_1(\mathbf{x}_1) = N \int \Psi(\mathbf{x}_1, \mathbf{x}_2, \ldots, \mathbf{x}_N)\, \Psi^*(\mathbf{x}_1, \mathbf{x}_2, \ldots, \mathbf{x}_N)\, d\mathbf{x}_2 \cdots d\mathbf{x}_N \qquad (4.1)$$

It should be noted that this statement implies a slight change in the interpretation of $\mathbf{x}_1$ as an argument in the density function; it no longer denotes exclusively the variables of particle 1 but rather a point of configuration space at which *any* particle may be found (with equal probability owing to indistinguishability). Thus $\rho_1(\mathbf{x}_1)\, d\mathbf{x}_1$ is the probability of finding any electron $(1, 2, \ldots, N)$ in volume element $d\mathbf{x}_1$ at $\mathbf{x}_1$.

As in the case of one electron, we can also obtain the probability of an electron in a volume element $d\mathbf{r}_1$ in ordinary three-dimensional space, with *any* spin value, by integrating over s_1. Thus

$$P_1(\mathbf{r}_1) = \int \rho_1(\mathbf{x}_1)\, ds_1 \qquad (4.2)$$

is the probability per unit volume of finding an electron at point $\mathbf{r}_1$. This is the ordinary *electron density function* measured by crystallographers.

We have put a subscript 1 on the density functions to indicate reference to *one* particle, but it is also possible to introduce probabilities for different configurations or "clusters" of any number of particles. Thus,

$$\rho_2(\mathbf{x}_1, \mathbf{x}_2) = N(N-1) \int \Psi(\mathbf{x}_1, \mathbf{x}_2, \ldots, \mathbf{x}_N)\, \Psi^*(\mathbf{x}_1, \mathbf{x}_2, \ldots, \mathbf{x}_N)\, d\mathbf{x}_3 \cdots d\mathbf{x}_N \qquad (4.3)$$

determines the probability of two electrons (*any* two) being found simultaneously at "points" $\mathbf{x}_1$, $\mathbf{x}_2$ (spin included), while

$$P_2(\mathbf{x}_1, \mathbf{x}_2) = \int \rho_2(\mathbf{x}_1, \mathbf{x}_2)\, ds_1\, ds_2 \qquad (4.4)$$

is the probability of finding them at $\mathbf{r}_1$ and $\mathbf{r}_2$ (in ordinary space) with *any* combination of spins (one up, one down; both up; both down). The function P_2 is a "pair function," exactly analogous to a corresponding function used in the theory of liquids. Because electrons interact only pairwise, there is no need to consider distribution functions referring to more than two at a time.

Before looking at the characteristic orbital forms of density functions such as P_1 and P_2, it is desirable to make a further slight generalization. To see why this is so, we again consider first a one-electron system, an electron in spin-orbital ψ. The expectation value in this state of any quantity with operator F is given by

$$\langle F \rangle = \int \psi^*(\mathbf{x})\, \mathsf{F}\psi(\mathbf{x})\, d\mathbf{x} \tag{4.5}$$

If F is just a multiplier (e.g., a function F of coordinates and spin), we can write

$$\langle F \rangle = \int F\psi(\mathbf{x})\, \psi^*(\mathbf{x})\, d\mathbf{x} = \int F\rho_1(\mathbf{x})\, d\mathbf{x}$$

since the order of the factors does not matter; the expectation value is thus obtained simply by averaging $F = F(\mathbf{x})$ over the electron density, since $\rho_1(\mathbf{x})\, d\mathbf{x}$ is the probability of finding an electron in $d\mathbf{x}$ with $F = F(\mathbf{x})$, and the integrand is the corresponding contribution to the ordinary classical average value of F. It would be nice to do this when F is a true *operator* (e.g., involving differentiation or integration), but this does not seem to be possible because the operator F will work on everything that follows it, and $\psi^*(\mathbf{x})$ cannot therefore be moved to the right of F. In order to express everything in terms of the basic density function, we use a very simple device; we agree that F works on functions of $\mathbf{x}$ only, and we change the name of the variable in ψ^* from $\mathbf{x}$ to $\mathbf{x}'$ to make it immune from the effect of F. We can then write the expectation value as

$$\langle F \rangle = \int_{\mathbf{x}'=\mathbf{x}} \mathsf{F}\psi(\mathbf{x})\, \psi^*(\mathbf{x}')\, d\mathbf{x}$$

where we put $\mathbf{x}' = \mathbf{x}$ *after* operating with F but before completing the integration. For one electron, $\psi(\mathbf{x})\, \psi^*(\mathbf{x}) = \rho_1(\mathbf{x})$; and we now denote $\psi(\mathbf{x})\, \psi^*(\mathbf{x}')$ by $\rho_1(\mathbf{x}; \mathbf{x}')$ but continue to use $\rho_1(\mathbf{x})$ for the function obtained by identifying the two sets of variables, $\mathbf{x}$ and $\mathbf{x}'$:

$$\rho_1(\mathbf{x}) = \rho_1(\mathbf{x}; \mathbf{x})$$

The same artifice may be used in the many-electron case. Instead of $\rho_1(\mathbf{x}_1)$, we define the generalized density

$$\rho_1(\mathbf{x}_1; \mathbf{x}_1') = N \int \Psi(\mathbf{x}_1, \mathbf{x}_2, ..., \mathbf{x}_N)\, \Psi^*(\mathbf{x}_1', \mathbf{x}_2, ..., \mathbf{x}_N)\, d\mathbf{x}_2 \cdots d\mathbf{x}_N \quad (4.6)$$

in which we have put a prime on the $\mathbf{x}_1$ in Ψ^*. This protects it from the action of an operator, as in the one-electron case, and allows us to discuss *all* one-electron properties in terms of the single density function $\rho_1(\mathbf{x}_1; \mathbf{x}_1')$.

Let us take, for example, the Hamiltonian for a many-electron system:

$$\mathsf{H} = \sum_i \mathsf{h}(i) + \tfrac{1}{2} \sum_{i,j}{}' g(i, j) \quad (4.7)$$

where $\mathsf{h}(i)$ is the one-electron Hamiltonian for electron i in the field of the nuclei, and the $g(i, j)$ are Coulomb interaction energies. The expectation value of the one-electron part of the energy is[2]

$$\Big\langle \sum_i \mathsf{h}(i) \Big\rangle = N \int \Psi^*(\mathbf{x}_1, \mathbf{x}_2, ..., \mathbf{x}_N)\, \mathsf{h}(1)\, \Psi(\mathbf{x}_1, \mathbf{x}_2, ..., \mathbf{x}_N)\, d\mathbf{x}_1 \cdots d\mathbf{x}_N$$

$$= N \int \mathsf{h}(1)\, \Psi(\mathbf{x}_1, \mathbf{x}_2, ..., \mathbf{x}_N)\, \Psi^*(\mathbf{x}_1', \mathbf{x}_2, ..., \mathbf{x}_N)\, d\mathbf{x}_1 \cdots d\mathbf{x}_N$$

[2] Read this as "expectation value of the quantity with operator $\sum_i \mathsf{h}(i)$."

and by comparison with the definition (4.6), this means

$$\left\langle \sum_i \mathsf{h}(i) \right\rangle = \int_{\mathbf{x}_1'=\mathbf{x}_1} \mathsf{h}(1)\, \rho_1(\mathbf{x}_1; \mathbf{x}_1')\, d\mathbf{x}_1 \tag{4.8}$$

The expectation value of a one-electron operator (or, more correctly, the symmetric *sum* of such operators) is thus related directly to the *one-electron* density function. The wave function itself, with all its superfluous variables has receded into the background. It is usual to call $\rho_1(\mathbf{x}_1; \mathbf{x}_1')$ the one-electron *density matrix*. Density matrices have been used for a long time in statistical mechanics [*2, 3*] (see also, e.g., Tolman [*4*], but their value is dealing with electronic systems was not fully appreciated until comparatively recently.

In an exactly similar way the electron interaction energy, which involves *pairs* of electrons, and hence *two*-electron operators, may be written in terms of the "pair" function $\rho_2(\mathbf{x}_1, \mathbf{x}_2; \mathbf{x}_1', \mathbf{x}_2')$:

$$\left\langle \sum_{i,j}{}' g(i,j) \right\rangle = \int_{\substack{\mathbf{x}_1'=\mathbf{x}_1 \\ \mathbf{x}_2'=\mathbf{x}_2}} g(1,2)\, \rho_2(\mathbf{x}_1, \mathbf{x}_2; \mathbf{x}_1', \mathbf{x}_2')\, d\mathbf{x}_1\, d\mathbf{x}_2 \tag{4.9}$$

Since only one-body and two-body interactions are present for electrons, *we never need any density functions beyond* $\rho_1(\mathbf{x}_1; \mathbf{x}_1')$ *and* $\rho_2(\mathbf{x}_1, \mathbf{x}_2; \mathbf{x}_1', \mathbf{x}_2')$! For operators like $g(i,j)$, of course, which are merely multiplicative factors, there is no need to distinguish primed and unprimed variables; we may simply use the "diagonal elements"

$$\rho_1(\mathbf{x}_1) = \rho_1(\mathbf{x}_1; \mathbf{x}_1), \qquad \rho_2(\mathbf{x}_1, \mathbf{x}_2) = \rho_2(\mathbf{x}_1, \mathbf{x}_2; \mathbf{x}_1, \mathbf{x}_2) \tag{4.10}$$

which are the density functions we started from in (4.1) and (4.3).

In dealing with spin-independent effects we can go further by completing the spin integrations over the remaining one or two spin variables. The analogs of the *spinless* densities introduced in (4.2) and (4.4) are

$$P_1(\mathbf{r}_1;\mathbf{r}_1') = \int_{s_1'=s_1} \rho_1(\mathbf{x}_1;\mathbf{x}_1')\,ds_1 \tag{4.11}$$

$$P_2(\mathbf{r}_1,\mathbf{r}_2;\mathbf{r}_1',\mathbf{r}_2') = \int_{\substack{s_1'=s_1\\ s_2'=s_2}} \rho_2(\mathbf{x}_1,\mathbf{x}_2;\mathbf{x}_1',\mathbf{x}_2')\,ds_1\,ds_2 \tag{4.12}$$

and on removing the primes we obtain the electron density $P_1(\mathbf{r}_1)$ and the pair function $P_2(\mathbf{r}_1, \mathbf{r}_2)$, exactly as defined originally in (4.2) and (4.4). These quantities are easy to visualize, whereas the wave function is not. It is well known, for example, that in an orbital approximation P_1 may be "divided out" by giving the *electron populations* of various well-defined orbital and overlap regions.

The conceptual value of the density functions becomes clear on writing down the energy expression for a molecule in the usual first approximation, spin terms in the Hamiltonian being neglected. The result is

$$\begin{aligned} E = &\int_{\mathbf{r}_1'=\mathbf{r}_1} -\tfrac{1}{2}\nabla^2(1)\,P_1(\mathbf{r}_1;\mathbf{r}_1')\,d\mathbf{r}_1 && \text{(kinetic energy)} \\ &+ \int V(1)\,P_1(\mathbf{r}_1)\,d\mathbf{r}_1 && \left(\begin{array}{l}\text{potential energy of smeared-}\\ \text{out charge, density } P_1\text{, in field}\\ \text{of nuclei}\end{array}\right) \\ &+ \tfrac{1}{2}\int g(1,2)\,P_2(\mathbf{r}_1,\mathbf{r}_2)\,d\mathbf{r}_1\,d\mathbf{r}_2 && \left(\begin{array}{l}\text{average energy due to pair-wise}\\ \text{repulsions of electrons with}\\ \text{pair function } P_2\end{array}\right) \end{aligned} \tag{4.13}$$

The first term is the quantum-mechanical expectation value of the kinetic energy, but the others have a purely classical interpretation in terms of the distribution functions for a particle and a pair of particles, respectively. The second term has an obvious pictorial significance, while the third is familiar from the theory of fluids (see, e.g., Born and Green [5]), the factor $\frac{1}{2}$

compensating for the double counting of each volume element when $\mathbf{r}_1$ and $\mathbf{r}_2$ independently run over all space. The "reduced density matrices" ρ_1 and ρ_2, etc., were introduced by Husimi [*6*], and the present normalization has also been followed by McWeeny [*1a–c*], Born and Green [*5*], Yang [*7*], and others. Alternative normalizations have been used by Löwdin [*8*], ter Haar [*9*], and Coleman [*10*].

The above results are valid for all kinds of wave functions, or approximate wave functions, for any state of any system; because they involve the electron distribution directly, it is often possible to get a useful interpretation of molecular properties knowing only the main features of the electron density and without reference to the intricacies of the many-electron wave function. A chemical bond, for instance, arises from a concentration of electron density in the bond region, with a consequent lowering of potential energy—giving clear support to the intuitive ideas of elementary valence theory.

Spin Density

The question that concerns us next is whether we can introduce a parallel, and equally convenient, description of the *spin* distribution in a molecule. We shall examine this possibility within the context of a simple example.

Let us consider a three-electron system, such as the lithium atom or the π-electron part of the allyl radical, using a one-determinant MO approximation with two electrons in the lowest MO, A say, and one in the next, B. The wave function is

$$\Psi = | A\alpha\, A\beta\, B\alpha | / \sqrt{3!}$$

and Slater's rules [*11*] tell us that the one-electron contribution to the energy is

$$\left\langle \Psi \middle| \sum_i \mathsf{h}(i) \middle| \Psi \right\rangle = \langle A\alpha \mid \mathsf{h} \mid A\alpha \rangle + \langle A\beta \mid \mathsf{h} \mid A\beta \rangle + \langle B\alpha \mid \mathsf{h} \mid B\alpha \rangle$$

where a typical one-electron matrix element is

$$\langle A\alpha \mid \mathsf{h} \mid A\alpha\rangle = \int A^*(\mathbf{r}_1)\, \alpha^*(s_1)\, \mathsf{h}(1)\, A(\mathbf{r}_1)\, \alpha(s_1)\, d\mathbf{r}_1\, ds_1$$

If we write this as

$$\int_{\substack{s_1'=s_1\\ \mathbf{r}_1'=\mathbf{r}_1}} \mathsf{h}(1)\, A(\mathbf{r}_1)\, \alpha(s_1)\, A^*(\mathbf{r}_1')\, \alpha^*(s_1')\, d\mathbf{r}_1\, ds_1$$

and recall that

$$\left\langle \Psi \,\middle|\, \sum_i \mathsf{h}(i) \,\middle|\, \Psi \right\rangle = \int_{\mathbf{x}_1'=\mathbf{x}_1} \mathsf{h}(1)\, \rho_1(\mathbf{x}_1; \mathbf{x}_1')\, d\mathbf{x}_1$$

it is clear that the matrix element $\langle A\alpha \mid \mathsf{h} \mid A\alpha\rangle$ is associated with a term $A(\mathbf{r}_1)\, \alpha(s_1)\, A^*(\mathbf{r}_1')\, \alpha^*(s_1')$ in the density matrix $\rho_1(\mathbf{x}_1; \mathbf{x}_1')$; putting all the terms together gives

$$\begin{aligned}\rho_1(\mathbf{x}_1; \mathbf{x}_1') = {} & [A(\mathbf{r}_1)\, A^*(\mathbf{r}_1') + B(\mathbf{r}_1)\, B^*(\mathbf{r}_1')]\, \alpha(s_1)\, \alpha^*(s_1') \\ & + [A(\mathbf{r}_1)\, A^*(\mathbf{r}_1')]\, \beta(s_1)\, \beta^*(s_1')\end{aligned}$$

This is a special case of the general result

$$\rho_1(\mathbf{x}_1; \mathbf{x}_1') = P_1^{\alpha,\alpha}(\mathbf{r}_1; \mathbf{r}_1')\, \alpha(s_1)\, \alpha^*(s_1') + P_1^{\beta,\beta}(\mathbf{r}_1; \mathbf{r}_1')\, \beta(s_1)\, \beta^*(s_1') \tag{4.14}$$

which expresses the density function in terms of components referring to up-spin and down-spin electrons separately. If we specialize to the "diagonal element" and integrate over spin to get the ordinary electron density, we obtain [with an abbreviated notation, like that used in (4.10)]

$$P_1(\mathbf{r}_1) = P_1^{\alpha}(\mathbf{r}_1) + P_1^{\beta}(\mathbf{r}_1) \tag{4.15}$$

where the parts arising from $s_1 \simeq +\frac{1}{2}$ and $s_1 \simeq -\frac{1}{2}$, respectively, represent the densities of up-spin and down-spin electrons. This is exactly what we expect, since the electron density associated with electrons in $A\alpha$ and $B\alpha$ is $|A|^2 + |B|^2$ (up-spin), while there is only one β-type spin-orbital $A\beta$, giving a down-spin density $|A|^2$. But the resolution into components is quite general; however many determinants the wave function may contain, and as long as Ψ is a state of definite spin (quantum numbers S and M), there are no "cross terms" of $\alpha\beta^*$ or $\beta\alpha^*$ type. For a one-determinant wave function, the situation is particularly simple, the up-spin and down-spin densities being sums of corresponding orbital contributions.

It is now natural to define a *spin density* as

$$Q_S(\mathbf{r}_1) = \tfrac{1}{2}(P_1^{\alpha}(\mathbf{r}_1) - P_1^{\beta}(\mathbf{r}_1))$$

since this measures the resultant z component of spin (excess of up-spin over down-spin density multiplied by the spin magnitude $\frac{1}{2}$). More generally, we write

$$Q_S(\mathbf{r}_1; \mathbf{r}_1') = \tfrac{1}{2}[P_1^{\alpha,\alpha}(\mathbf{r}_1; \mathbf{r}_1') - P_1^{\beta,\beta}(\mathbf{r}_1; \mathbf{r}_1')] \tag{4.16}$$

This definition is clearly appropriate, since if we write down the density matrix expression for the expectation value of total spin z component S_z, we obtain

$$\left\langle \Psi \middle| \sum_i \mathsf{S}_z(i) \middle| \Psi \right\rangle = \int_{\mathbf{x}_1'=\mathbf{x}_1} \mathsf{S}_z(1)\, \rho_1(\mathbf{x}_1; \mathbf{x}_1')\, d\mathbf{x}_1$$

$$= \int_{\substack{\mathbf{r}_1'=\mathbf{r}_1 \\ s_1'=s_1}} [P_1^{\alpha,\alpha}(\mathbf{r}_1; \mathbf{r}_1')\, \alpha(s_1)\, \alpha^*(s_1')$$

$$- P_1^{\beta,\beta}(\mathbf{r}_1; \mathbf{r}_1')\, \beta(s_1)\, \beta^*(s_1')]\, d\mathbf{r}_1\, ds_1$$

and hence, from (4.16),

$$\langle \Psi | \mathsf{S}_z | \Psi \rangle = \int Q_S(\mathbf{r}_1)\, d\mathbf{r}_1 = M \tag{4.17}$$

just as if the z component of spin angular momentum were spread out over the molecule with a density $Q_S(\mathbf{r}_1)$, integration over all space yielding the total expectation value which, for a definite spin state (with quantum numbers S, M), must be M (in units of $\hbar = h/2\pi$). We could normalize this density in various ways; if we omitted the factor $\frac{1}{2}$ in the definition (4.16), integration would yield $N_\alpha - N_\beta$, the number of "free" spins; if we divided by M, (4.17) shows that integration would yield unity, and the density

$$D_S(\mathbf{r}_1; \mathbf{r}_1') = Q_S(\mathbf{r}_1; \mathbf{r}_1')/M \tag{4.18}$$

would have a diagonal element coincident with the "normalized spin density" defined by McConnell [*12a*] (see also Weissman [*12b*]). It is clear that Q_S arises from *off*-diagonal elements ($\mathbf{x}_1' \neq \mathbf{x}_1$) of ρ_1, since S_z has operator properties; and we shall find that off-diagonal elements of Q_S itself ($\mathbf{r}_1' \neq \mathbf{r}_1$) are important when we wish to relate expectation values of spin interaction terms to the spin density. It is also clear from the derivation of (4.17) that the spin density matrix may be written in the form

$$Q_S(\mathbf{r}_1; \mathbf{r}_1') = \int_{s_1'=s_1} S_z(1)\, \rho_1(\mathbf{x}_1; \mathbf{x}_1')\, ds_1 \tag{4.19}$$

as an alternative to (4.16).

Finally we note [*13*] that the actual density of spin angular momentum has the same functional form in each state of a spin multiplet ($M = S, S - 1, \ldots, -S$), although its magnitude at any given point is proportional to M. Thus we can write

$$Q_S(\mathbf{r}_1; \mathbf{r}_1') = (M/S)\, Q_S(\mathbf{r}_1; \mathbf{r}_1')_{st} \tag{4.20}$$

where $Q_S(\mathbf{r}_1; \mathbf{r}_1')_{st}$ is the spin density in the "standard state" with $M = S$; the spin density need thus be calculated only for *one state* of any given multiplet.

At this point, it is perhaps worth illustrating the conceptual value of the spin density matrix by quoting two results, whose derivation we shall discuss later. For rapidly tumbling molecules in a liquid, there is an observed electron-nuclear coupling—

arising from the "contact term" in the full Hamiltonian [see (3.19)]—represented by a spin-Hamiltonian term containing both electron (S) and nuclear (I) spin operators:

$$\mathsf{H}_S(\text{cont}) = \sum_n a_n^{\text{cont}} \mathbf{S} \cdot \mathbf{I}(n)$$

The coupling constant a_n^{cont} depends on the electron spin density evaluated at $\mathbf{R}_n$, the position of nucleus n. Using β_{p} for the *nuclear* magneton (proton mass M instead of m), it is

$$a_n^{\text{cont}} = (8\pi g\beta g_n\beta_{\text{p}}/3)\, D_S(\mathbf{R}_n)$$

In a crystal, on the other hand, there is a *tensor* coupling arising from the dipole-dipole terms in the Hamiltonian, and this no longer averages to zero. It can be represented by a spin-Hamiltonian term

$$\mathsf{H}_S(\text{dip}) = \sum_n \sum_{\lambda,\mu} a_{n,\lambda\mu}^{\text{dip}} \mathsf{S}_\lambda \mathsf{I}_\mu(n) \qquad (\lambda, \mu = x, y, z)$$

and the coupling tensor has elements depending on the second moments of the spin density about nucleus n, e.g.,

$$a_{n,xy}^{\text{dip}} = (3g\beta g_n\beta_{\text{p}}) \int (x_n y_n/r_n^5)\, D_S(\mathbf{r})\, d\mathbf{r}$$

where $\mathbf{r}_n$ is the distance of the field point ($\mathbf{r}$) from nucleus n (and has Cartesian components x_n, y_n, z_n). The spin density is therefore directly related to *observables*; Its significance in ESR and NMR spectroscopy is comparable with that of the electron density in X-ray crystallography.

Population Analysis

We now turn to the numerical characterization of the spin density function, assuming that the wave function from which it is derived is constructed, in the last analysis, from some given basis of atomic orbitals ϕ_1, ϕ_2,..., ϕ_i,... . This does not prejudge the theoretical method to be used; thus, we might construct VB-type wave functions, using the ϕ's directly, or we might use an MO function with an LCAO approximation to each MO.

It is well known that, irrespective of the approach used, the electron density takes the form

$$P_1(\mathbf{r}) = \sum_{i,j} P_{1ij}\phi_i(\mathbf{r})\,\phi_j{}^*(\mathbf{r}) \tag{4.21}$$

and that a statement of the numerical values of the coefficients P_{1ij} therefore completely describes the distribution of charge over the molecule. This observation is the basis of "population analysis"; we define normalized orbital and overlap densities (assuming for convenient real orbitals)

$$d_i(\mathbf{r}) = \phi_i(\mathbf{r})^2, \qquad d_{ij}(\mathbf{r}) = \phi_i(\mathbf{r})\,\phi_j(\mathbf{r})/S_{ij} \qquad (i \neq j) \tag{4.22}$$

where S_{ij} is an overlap integral, and rewrite the charge density as

$$P_1(\mathbf{r}) = \sum_i q_i d_i(\mathbf{r}) + \sum_{i<j} q_{ij} d_{ij}(\mathbf{r}) \tag{4.23}$$

where

$$q_i = P_{1ii}\,, \qquad q_{ij} = 2S_{ij}P_{1ij} \qquad (i \neq j) \tag{4.24}$$

Now since integration of the charge density over all space must yield the total number of electrons N, it is clear that

$$\int P_1(\mathbf{r})\,d\mathbf{r} = \sum_i q_i + \sum_{i<j} q_{ij} = N \tag{4.25}$$

and that the q's describe how the N electrons are divided among the various orbital and overlap regions. The q_i and q_{ij}, originally used in π-electron theory [*14*] as atom and bond "charges" (related to the Coulson charges and bond orders) and applied also in crystallography [*15–17*] are nowadays referred to as orbital and overlap "populations." The results of an early calculation [*17*] of orbital and overlap populations, for a carbon atom in graphite, are shown in Fig. 4.1. Clearly, by

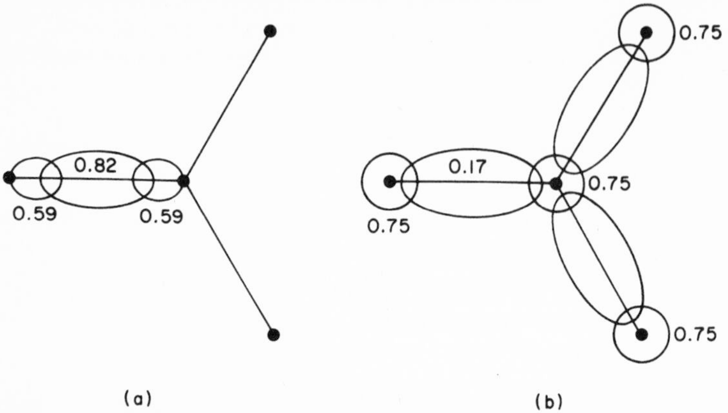

FIG. 4.1. Electron populations for carbon in graphite. (a) Typical σ bond (hybrid orbital and overlap regions indicated schematically). (b) Three partial π bonds. The numbers give the populations (in electrons) of the various regions.

summing the orbital charges on an atom, and then adding half of every overlap charge connecting it with neighboring atoms, one obtains a "gross charge," which may be formally associated with the atom in its molecular environment. The corresponding distribution was used in calculating an "effective" X-ray scattering factor for a bonded carbon atom in graphite. Population analysis has been extensively developed by Mulliken [*18*] and is widely used in describing and comparing the results of wave-function calculations.

Spin populations are defined in an exactly similar way. If we use the normalized spin density (4.18), we may write

$$D_S(\mathbf{r}) = \sum_{ij} D_{Sij}\phi_i(\mathbf{r})\,\phi_j^*(\mathbf{r}) \tag{4.26}$$

and introduce spin populations for the orbital and overlap regions by

$$q_i^{\mathrm{S}} = D_{\mathrm{S}ii}, \qquad q_{ij}^{\mathrm{S}} = 2S_{ij}D_{\mathrm{S}ij} \tag{4.27}$$

Since integration over all space must in this case yield unity, the "conservation equation" for spin density is [cf. (4.25)]

$$\sum_i q_i^{\mathrm{S}} + \sum_{i<j} q_{ij}^{\mathrm{S}} = 1 \tag{4.28}$$

It should be noted that with *orthogonal* basis orbitals the overlap populations vanish, for both charge and spin. In this case, for example, we obtain $\sum_i q_i^{\mathrm{S}} = 1$, and this condition is frequently invoked in fixing the signs of spin populations when these are not given unequivocally by experiment. Such a procedure is, however, precarious. What is usually inferred from observation is not a population but the spin *density* around a nucleus (p. 80).

It is interesting to look at some representative calculations of spin density for a simple π-electron system, the allyl radical, to compare the convergence of different approaches. These recent calculations were made [*19*] using a basis of three (orthogonalized) 2p AO's, from which four independent 3-electron functions of $^2A_2$ symmetry can be set up—either symmetry combinations of VB structures or MO functions arising from suitable excitations from a one-determinant SCF function. The results are shown in Fig. 4.2 for three levels of approximation.

What is most striking is the ability of the simplest VB function to predict correctly a negative spin density on the central atom, in agreement with experiment and the full CI calculation, even when the wave function is so poor in its *orbital* form that it gives zero bond orders and a total π-electron energy in error by nearly 20%. Similar results for the first triplet state of butadiene appear in Fig. 4.3, a VB calculation with no polar structures (and no π bonding) correctly reproducing the main features of the spin distribution. Presumably, this is because the energy is heavily dependent on overlap of the AO's, while the spin distribution is determined mainly by the available coupling

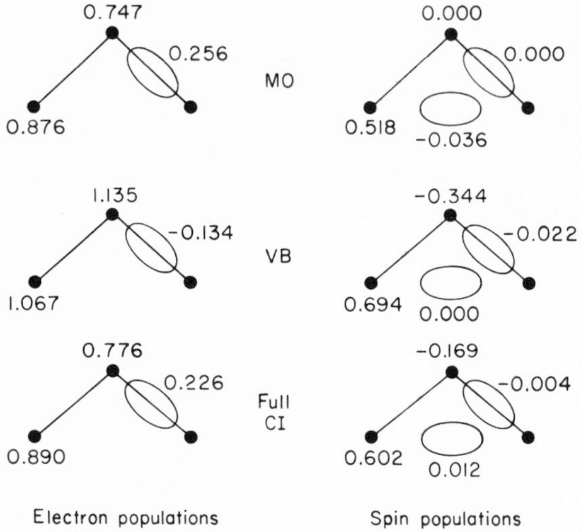

FIG. 4.2. Electron and spin populations in the allyl radical (π electrons only). Orbital populations are indicated by numbers at the vertices, while overlap populations (where significant) are indicated by those in the overlap regions (schematic). MO and VB approximations converge to the full CI results as "excited functions" or "polar structures" are added. Populations not shown are evident from symmetry.

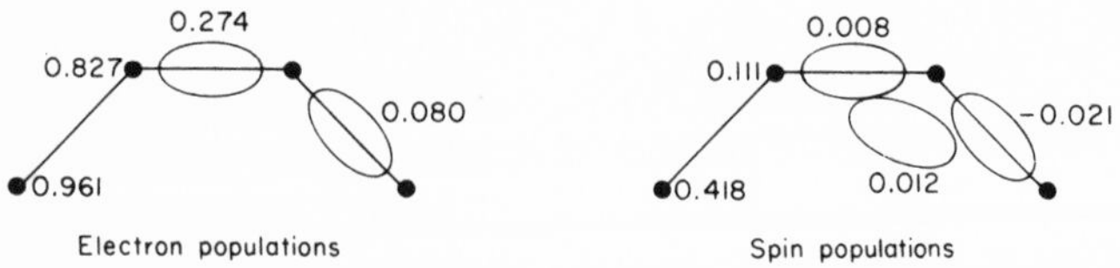

FIG. 4.3. Electron and spin populations in cis-butadiene triplet state (π electrons only). The results arise from full CI π-electron calculations. Populations are indicated as in Fig. 4.2.

schemes in a few principal structures. With orthogonal AO's the actual π bonding is introduced via the *polar* structures, but the weights with which these are brought in is not sufficient to change the main features of the coupling.

SPIN CORRELATION

The charge and spin densities together fully determine ρ_1 and hence the expectation values of all *one*-electron quantities. Some effects, however, depend on two electrons at a time and on the spins of one or both of them. An example is the direct electron spin-spin coupling due to dipolar interactions, an effect already recognized in the H_{SS} term (3.18) of the full Hamiltonian. To deal in a general way with such operators, we need to resolve ρ_2 into components corresponding to different spin situations, just as we did in the case of ρ_1. Thus we might write

$$\begin{aligned}\rho_2(\mathbf{x}_1, \mathbf{x}_2; \mathbf{x}_1', \mathbf{x}_2') = {} & P_2^{\alpha\alpha,\alpha\alpha}(\mathbf{r}_1, \mathbf{r}_2; \mathbf{r}_1', \mathbf{r}_2')\, \alpha(s_1)\, \alpha(s_2)\, \alpha^*(s_1')\, \alpha^*(s_2') \\ & + P_2^{\alpha\alpha,\alpha\beta}(\mathbf{r}_1, \mathbf{r}_2; \mathbf{r}_1', \mathbf{r}_2')\, \alpha(s_1)\, \alpha(s_2)\, \alpha^*(s_1')\, \beta^*(s_2') \\ & + \cdots \qquad (4.29)\end{aligned}$$

where there would appear to be 16 distinct components. If we consider the diagonal element ($\mathbf{x}_1' = \mathbf{x}_1$, $\mathbf{x}_2' = \mathbf{x}_2$) and integrate over all spin possibilities to obtain $P_2(\mathbf{r}_1, \mathbf{r}_2)$, only four contributions remain:

$$P_2(\mathbf{r}_1, \mathbf{r}_2) = P_2^{\alpha\alpha}(\mathbf{r}_1, \mathbf{r}_2) + P_2^{\alpha\beta}(\mathbf{r}_1, \mathbf{r}_2) + P_2^{\beta\alpha}(\mathbf{r}_1, \mathbf{r}_2) + P_2^{\beta\beta}(\mathbf{r}_1, \mathbf{r}_2)$$

where each term has a clear physical meaning. Thus

$$P_2^{\alpha\beta}(\mathbf{r}_1, \mathbf{r}_2)\, d\mathbf{r}_1\, d\mathbf{r}_2$$

is the probability of finding a spin-up electron in $d\mathbf{r}_1$ at $\mathbf{r}_1$ at the same time as a spin-down electron in $d\mathbf{r}_2$ at $\mathbf{r}_2$. Such functions [*1a*] thus allow us to describe the *correlation* between the spins of electrons at different points. Again, for generality, we need the off-diagonal elements such as $P_2^{\alpha\beta,\alpha\beta}(\mathbf{r}_1, \mathbf{r}_2; \mathbf{r}_1', \mathbf{r}_2')$ and we find that for a definite spin state just *six* of the 16 components may be nonzero. The extra terms, $P_2^{\alpha\beta,\beta\alpha}$ and $P_2^{\beta\alpha,\alpha\beta}$, are needed when we require expectation values of operators such as $\mathbf{S}(1) \cdot \mathbf{S}(2)$ which can switch two spins.

As in the case of the spin density, it is not the individual components of ρ_2 that are of most interest, but rather certain linear combinations of them. How, for example, could we measure the *anisotropy* in the spin distribution due to the imposition of a powerful magnetic field? Since in an isotropic situation the expectation values of all three components of $\mathbf{S}^2 = S_x{}^2 + S_y{}^2 + S_z{}^2$ would be equal, it seems reasonable to measure the "magnetic polarization" around the z direction by the expectation values of $3S_z{}^2 - \mathbf{S}^2$. In a definite spin state, with M fixing the spin component along the z axis, this expectation value takes the value $3M^2 - S(S+1)$, whereas in an isotropic situation it would vanish. Now

$$3S_z{}^2 - \mathbf{S}^2 = \sum_{i,j}{}' [3S_z(i)\, S_z(j) - \mathbf{S}(i) \cdot \mathbf{S}(j)] \tag{4.30}$$

where we can omit the term $i = j$ (prime on the summation sign) because it is zero anyway, each part being equivalent to multiplication by $\frac{3}{4}$ (see p. 11)

$$\langle \Psi \mid 3S_z{}^2 - \mathbf{S}^2 \mid \Psi \rangle = \int_{\mathbf{x}_1'=\mathbf{x}_1} [3S_z(1)\, S_z(2) - \mathbf{S}(1) \cdot \mathbf{S}(2)]$$

$$\times\ \rho_2(\mathbf{x}_1, \mathbf{x}_2; \mathbf{x}_1', \mathbf{x}_2')\, d\mathbf{x}_1\, d\mathbf{x}_2$$

and obtain, after performing the spin integrations

$$\int Q_{SS}(\mathbf{r}_1, \mathbf{r}_2)\, d\mathbf{r}_1\, d\mathbf{r}_2 = 3M^2 - S(S+1) \tag{4.31}$$

where $Q_{SS}(\mathbf{r}_1, \mathbf{r}_2)$, with the usual abbreviation, denotes a diagonal element of the function

$$Q_{SS}(\mathbf{r}_1, \mathbf{r}_2; \mathbf{r}_1', \mathbf{r}_2') = \int_{\substack{s_1'=s_1 \\ s_2'=s_2}} [3S_z(1)\, S_z(2) - \mathbf{S}(1) \cdot \mathbf{S}(2)]$$

$$\times\, \rho_2(\mathbf{x}_1, \mathbf{x}_2; , \mathbf{x}_1' \mathbf{x}_2')\, ds_1\, ds_2 \qquad (4.32)$$

Physically, the contributions to the anisotropy, measured by $3M^2 - S(S+1)$, come from all configurations of a pair of electrons; the little bit that arises from electrons in volume elements $d\mathbf{r}_1$ and $d\mathbf{r}_2$, at points $\mathbf{r}_1$ and $\mathbf{r}_2$, is $Q_{SS}(\mathbf{r}_1, \mathbf{r}_2)\, d\mathbf{r}_1\, d\mathbf{r}_2$, and this quantity therefore measures the anisotropy in the coupling of the spins of electrons simultaneously occupying two different volume elements in space. The spin-spin dipolar coupling, which splits molecular Zeeman levels even in the zero-field limit (as was first observed by Hutchison and Mangum [*20*]), is completely determined by this one function—hence the subscript SS for the spin-spin coupling function. If we want to see how this looks in terms of a spin-up/spin-down description, we need only write ρ_2 in component form and examine the effect of the spin operator in the integral (4.32) that defines Q_{SS}. The result is

$$Q_{SS} = \tfrac{1}{2}\,[P_2^{\alpha\alpha,\alpha\alpha} - P_2^{\alpha\beta,\alpha\beta} - P_2^{\beta\alpha,\beta\alpha} + P_2^{\beta\beta,\beta\beta}] - \tfrac{1}{2}\,[P_2^{\alpha\beta,\beta\alpha} + P_2^{\beta\alpha,\alpha\beta}] \qquad (4.33)$$

Thus $Q_{SS}(\mathbf{r}_1, \mathbf{r}_2)$ depends on the difference of the probabilities of parallel and antiparallel spin situations, supplemented by two terms with no direct statistical meaning; the former determine matrix elements involving only spin z components, while the latter determine matrix elements involving components perpendicular to the axis of quantization.

It is well known (e.g., Dirac [*21*]) that in the usual Hartree–Fock approximation (one-determinant wave function), ρ_2 is completely determined by ρ_1. The spin-spin coupling function is reduced easily in this case, the result being

$$Q_{SS}(\mathbf{r}_1, \mathbf{r}_2; \mathbf{r}_1', \mathbf{r}_2') = 2[Q_S(\mathbf{r}_1; \mathbf{r}_1')\, Q_S(\mathbf{r}_2; \mathbf{r}_2') - Q_S(\mathbf{r}_2; \mathbf{r}_1')\, Q_S(\mathbf{r}_1; \mathbf{r}_2')] \tag{4.34}$$

In this approximation, therefore, the spin-spin coupling is determined entirely by the spin density, provided we use the generalized function with appropriate *off*-diagonal elements. The off-diagonal elements are analogous to similar terms in the pair function (P_2) and arise in a similar way from the Fermi correlation between electrons of the same spin.

As in the case of the spin density, the function Q_{SS} has the same functional form for all the states of a given spin multiplet. Since integration yields an expectation value $3M^2 - S(S+1)$, according to (4.31), the function may be normalized to unity by dividing by this factor:

$$D_{SS}(\mathbf{r}_1, \mathbf{r}_2; \mathbf{r}_1', \mathbf{r}_2') = Q_{SS}(\mathbf{r}_1, \mathbf{r}_2; \mathbf{r}_1', \mathbf{r}_2')/[3M^2 - S(S+1)] \tag{4.35}$$

Also in terms of the standard state with $M = S$, we obtain for any state of the multiplet

$$Q_{SS}(\mathbf{r}_1, \mathbf{r}_2; \mathbf{r}_1', \mathbf{r}_2') = \frac{3M^2 - S(S+1)}{S(2S-1)}\, Q_{SS}(\mathbf{r}_1, \mathbf{r}_2; \mathbf{r}_1', \mathbf{r}_2')_{st} \tag{4.36}$$

which is a useful result because the states with $M \neq S$ are not representable by single determinants and are consequently less easy to handle.

Only one other density function need be mentioned. It determines all spin-*orbit* coupling effects (including "spin-other-orbit" coupling, which is usually ignored) and may accordingly be denoted by $Q_{SL}(\mathbf{r}_1, \mathbf{r}_2; \mathbf{r}_1', \mathbf{r}_2')$. The definition is similar to that of Q_S in (4.19) but involves ρ_2 instead of ρ_1:

$$Q_{SL}(\mathbf{r}_1, \mathbf{r}_2; \mathbf{r}_1', \mathbf{r}_2') = \int_{\substack{s_1'=s_1 \\ s_2'=s_2}} \mathsf{S}_z(1)\, \rho_2(\mathbf{x}_1, \mathbf{x}_2; \mathbf{x}_1', \mathbf{x}_2')\, ds_1\, ds_2 \tag{4.37}$$

The diagonal element has a particularly straightforward physical .erpretation:

$$Q_{SL} = P_2^{\alpha\alpha} - P_2^{\beta\alpha} + P_2^{\alpha\beta} - P_2^{\beta\beta} \tag{4.38}$$

and is therefore proportional to the excess probability of finding a spin-up electron at $\mathbf{r}_1$, rather than a spin-down electron, when a second electron (either spin) is at $\mathbf{r}_2$. It is therefore a "conditional" spin density. Like the spin density (which may be obtained from this function by integration over $\mathbf{r}_2$), this function depends on M, within a multiplet, according to

$$Q_{SL}(\mathbf{r}_1, \mathbf{r}_2; \mathbf{r}_1', \mathbf{r}_2') = (M/S)Q_{SL}(\mathbf{r}_1, \mathbf{r}_2; \mathbf{r}_1', \mathbf{r}_2')_{st} \tag{4.39}$$

where the standard density again refers to the state with $M = S$. And again if we wish, we may introduce a "normalized" function $D_{SL} = Q_{SL}/M$ analogous to D_S in (4.18).

Together, the functions we have introduced determine the expectation values of all spin-dependent operators. They therefore determine, to first order of perturbation theory, the effect on the energy levels of any given molecular system of all the spin terms in the Hamiltonian. In the two remaining lectures we shall first examine some typical first-order effects, and then extend the analysis to second-order effects, trying to build a bridge between the density functions and the observable coupling constants in the molecular spin Hamiltonian.

REFERENCES

1a. McWeeny, R., *Proc. Roy. Soc. (London) Ser. A* **223**, 63 (1954).
1b. McWeeny, R., *Proc. Roy. Soc. (London) Ser. A* **232**, 114 (1955).
1c. McWeeny, R., *Rev. Mod. Phys.* **32**, 335 (1960).
2. von Neumann, J., *Nachr. Akad. Wiss. Goettingen Math. Physik. Kl. IIa* **1927**, 245 (1967).
3. Dirac, P. A. M., *Proc. Cambridge Phil. Soc.* **25**, 62 (1929).
4. Tolman, R. C., "The Principles of Statistical Mechanics." Oxford Univ. Press, London and New York, 1939.

5. Born, M., and Green, H. S., *Proc. Roy. Soc. (London) Ser. A* **191**, 168 (1947). *See also*, Born, M., "Natural Philosophy of Cause and Chance." Dover, New York, 1964.
6. Husimi, K., *Proc. Phys. Math. Soc. Japan* **22**, 264 (1940).
7. Yang, C. N., *Rev. Mod. Phys.* **34**, 694 (1962).
8. Löwdin, P.-O., *Phys. Rev.* **97**, 1474 (1955).
9. ter Haar, D., *Rept. Progr. Phys.* **24**, 304 (1961).
10. Coleman, A. J., *Rev. Mod. Phys.* **35**, 668 (1963).
11. Slater, J. C., *Phys. Rev.* **34**, 1293 (1929).
12a. McConnell, H. M., *J. Chem. Phys.* **28**, 1188 (1958).
12b. Weissman, S. I., *J. Chem. Phys.* **25**, 890 (1956).
13. McWeeny, R., and Mizuno, Y., *Proc. Roy. Soc. (London) Ser. A* **259**, 554 (1961).
14. McWeeny, R., *J. Chem. Phys.* **19**, 1614 (1951). *See also* Errata, *J. Chem. Phys.* **20**, 920 (1951).
15. McWeeny, R., *Acta Cryst.* **5**, 463 (1952).
16. McWeeny, R., *Acta Cryst.* **6**, 631 (1953).
17. McWeeny, R., *Acta Cryst.* **7**, 180 (1954).
18. Mulliken, R. S., *J. Chem. Phys.* **23**, 1833, 2343 (1955).
19. Cooper, I. L., and McWeeny, R., *J. Chem. Phys.* **49**, 3223 (1968).
20. Hutchison, C. A., and Mangum, B. W., *J. Chem. Phys.* **29**, 952 (1958).
21. Dirac, P. A. M., *Proc. Cambridge Phil. Soc.* **26**, 376 (1930).

5

NUCLEAR HYPERFINE EFFECTS AND ELECTRON SPIN-SPIN COUPLING

In the last two lectures we have looked at the spin terms in the *actual* Hamiltonian for a system of electrons and nuclei; we have considered the possibility of accounting for observed energy levels (and the way they are resolved by a magnetic field) in terms of a *fictitious* "spin Hamiltonian"; and we have discovered certain useful density functions that enable us to describe the distribution of charge and spin without always having to go back to the wave function. If the experimentalist gives us a phenomenological spin Hamiltonian, we can feel fairly sure that it contains much useful information about the form of the electron distribution. The question now is how to justify the spin-Hamiltonian concept, in a general way, and how to relate the numerical parameters that occur to the appropriate density functions. In trying to answer this question we follow the approach of McWeeny and Mizuno [*1*] and McWeeny [*2*] (following the normalization conventions of the latter). The use of the spin Hamiltonian in ESR has a long history, a key paper being that by Pryce [*3*]. The earlier work is covered by many excellent review papers (see, e.g., Bleaney and Stevens [*4*] and Low [*5*], as well as the book by Griffith [*6*]), but it is only more recently that attention has been turned to the detailed interpretation of parameters in terms of density functions.

A Simple Spin Hamiltonian

Let us first return to the simple illustration of a one-electron, one-nucleus system used in Lecture 1. Taking the nucleus to be a proton ($I = \frac{1}{2}$), and introducing a nuclear magneton (β_p) by using the proton mass M in place of the electronic m, the ESR spectrum can be fitted to a spin Hamiltonian of the form (field along the z axis)

$$\mathsf{H}_S = \underset{\text{(electronic Zeeman)}}{g\beta B \mathsf{S}_z} - \underset{\text{(nuclear Zeeman)}}{g_n\beta_p B \mathsf{I}_z} + \underset{\text{(hyperfine interaction)}}{a\mathbf{S}\cdot\mathbf{I}}$$

in the sense that if we take a set of electron-nuclear spin products $\alpha\alpha_n$, $\alpha\beta_n$, $\beta\alpha_n$, $\beta\beta_n$, expand the spin state as a linear combination, and solve the corresponding secular problem, then we get a perfect fit with experiment for some empirical value of the hyperfine coupling constant a. The secular determinant is, in fact, with $\Delta = g\beta B$, $\Delta_n = g_n\beta_p B$ and

$$X^{\pm} = \pm\tfrac{1}{2}(\Delta - \Delta_n) + \tfrac{1}{4}a, \qquad Y^{\pm} = \pm\tfrac{1}{2}(\Delta + \Delta_n) - \tfrac{1}{4}a$$

given by

$$\begin{array}{c} \alpha\alpha_n \\ \alpha\beta_n \\ \beta\alpha_n \\ \beta\beta_n \end{array}\begin{vmatrix} X^+ - E & 0 & 0 & 0 \\ 0 & Y^+ - E & \tfrac{1}{2}a & 0 \\ 0 & \tfrac{1}{2}a & Y^- - E & 0 \\ 0 & 0 & 0 & X^- - E \end{vmatrix} = 0$$

In the first approximation (a small) the levels are given by the diagonal elements alone, and these are the levels shown in the diagram of Fig. 1.6, which shows the effect on the single ESR signal (left-hand arrow) of admitting first Δ_n and then a. The off-diagonal elements show that the hyperfine interaction produces a slight mixing of the $\alpha\beta_n$ and $\beta\alpha_n$ states, which is easily taken into account to give a better approximation. The

Hamiltonian makes no reference to orbital motion, which is apparently relevant only in fixing the numerical value of the constant a.

General Theory

More generally, we may have an electronic ground state consisting of p_{e} degenerate levels, and there may be several nuclei, giving p_{n} possible nuclear spin-product functions (e.g., $\Theta_1^{\mathrm{nuc}} = \alpha(n)\,\alpha(n') \cdots$, $\Theta_2^{\mathrm{nuc}} = \alpha(n)\,\beta(n') \cdots$, etc.). The spin Hamiltonian assumed might then be

$$\mathsf{H}_{\mathrm{S}} = g\beta B\mathsf{S}_z - \sum_n g_n\beta_{\mathrm{p}}B\mathsf{I}_z(n) + \sum_n a_n\mathbf{S}\cdot\mathbf{I}(n) \qquad (5.1)$$

and to describe the splitting of the $p_{\mathrm{e}} \times p_{\mathrm{n}}$ degenerate levels, we should use electron-nuclear spin products $\Theta_M^{\mathrm{el}}\Theta_\lambda^{\mathrm{nuc}}$, where $\Theta_1^{\mathrm{el}}, \Theta_2^{\mathrm{el}}, \ldots,$ are $2S + 1$ spin functions ($M = S, S - 1, \ldots, -S$) corresponding to an "effective spin" S such that $p_{\mathrm{e}} = 2S + 1$. Usually, S coincides with the actual spin of the state considered (any *orbital* degeneracy being resolved, e.g., by Jahn–Teller distortion) but this is not essential (cf. p. 55), and since H_{S} anyway describes a fictitious "spin-only" system, S is frequently referred to as the "fictitious" spin.

It may be necessary to add other terms to (5.1), such as

$$\sum_{\lambda,\mu} D_{\lambda\mu}\mathsf{S}_\lambda\mathsf{S}_\mu \qquad (\lambda, \mu = x, y, z)$$

which describes an electron spin-spin coupling, and to introduce more flexibility into the Zeeman term by admitting a tensor coupling

$$\beta \sum_{\lambda,\mu} g_{\lambda\mu}B_\lambda\mathsf{S}_\mu$$

(the components referring to axes fixed in the molecule) which may differ greatly from the spin-only form $g\beta B\mathsf{S}_z$ with $g = 2$.

There will also be other nuclear terms of importance in NMR experiments, but we omit these for the moment.

We now recall the main features of a complete calculation, starting from the full Hamiltonian with its many small terms and formulating an appropriate secular problem. There is no escape now from the complexities of orbital motion, and we must expand the wave function in the form

$$\Psi = \sum_{\kappa\lambda} c_{\kappa\lambda} \Phi_\kappa \Theta_\lambda \tag{5.2}$$

When nuclear spins are taken into account the secular equations are

$$\mathbf{Hc} = E\mathbf{c} \tag{5.3}$$

where each row and column is indicated by a pair of labels (κ, λ). Thus, with an obvious abbreviation, $H_{\kappa'\lambda',\kappa\lambda} = \langle \kappa'\lambda' \mid \mathsf{H} \mid \kappa\lambda \rangle$ is the typical matrix element. We may again partition the equation as in Lecture 3 (p. 58) by dividing the $\{\Phi_\kappa\}$ into $\{\Phi_{\kappa_a}\}$ (A group) and $\{\Phi_{\kappa_b}\}$ (B group) and allowing all possible nuclear spin factors within each group. The blocks $\mathbf{H}^{AA}$, $\mathbf{H}^{AB}$, and $\mathbf{H}^{BB}$ then have elements:

$$\mathbf{H}^{AA} : \{\langle \kappa_a'\lambda' \mid H \mid \kappa_a\lambda \rangle\}$$

$$\mathbf{H}^{AB} : \{\langle \kappa_a'\lambda' \mid H \mid \kappa_b\lambda \rangle\}$$

$$\mathbf{H}^{BB} : \{\langle \kappa_b'\lambda' \mid H \mid \kappa_b\lambda \rangle\}$$

The A group comprises $p_{\mathrm{e}} \times p_{\mathrm{n}}$ functions, which are degenerate in the absence of spin-dependent and field-dependent terms in the Hamiltonian; it is this degeneracy that we must resolve. For simplicity, we assume in all that follows that only *spin* degeneracy is present in the A-group states (the commonest case). The presence of the B group makes itself felt by intergroup mixing, and we know that this can be taken into account by defining an *effective* Hamiltonian $\mathbf{H}_{\mathrm{eff}}$, essentially as in (3.25), whose leading terms are the $\mathbf{H}^{AA}$ block, followed by terms of second order in the off-diagonal elements. In full, putting $\mathsf{H} = \mathsf{H}_0 + \mathsf{H}'$, where the Φ_κ are approximate eigenfunctions of H_0 and H' contains the small terms,

$$\langle \kappa_a'\lambda' \mid \mathsf{H}_{\mathrm{eff}} \mid \kappa_a\lambda\rangle = E_a\, \delta_{\kappa_a'\kappa_a}\, \delta_{\lambda'\lambda} + \langle \kappa_a'\lambda' \mid \mathsf{H}' \mid \kappa_a\lambda\rangle + \sum_{\kappa_b''\lambda''} \frac{\langle \kappa_a'\lambda' \mid \mathsf{H}' \mid \kappa_b''\lambda''\rangle\langle \kappa_b''\lambda'' \mid \mathsf{H}' \mid \kappa_a\lambda\rangle}{(E_a - E_{\kappa_b''})} + \cdots \tag{5.4}$$

If all the energies are measured from E_a, the energy of the unresolved multiplet a, the first term (appearing only on the diagonal) may be dropped, while the next two give what we shall refer to as the "first-order and second-order" effects, respectively. This is merely a convenient classification, allowing us to give an unambiguous description of spin-Hamiltonian terms and the way in which they originate.

It will now be appreciated that finding a *first*-order spin Hamiltonian amounts to setting up an operator $\mathsf{H}_{\mathrm{S}}^{(1)}$, containing only spin operators and parameters, whose matrix elements between pure spin functions $\Theta_M^{\mathrm{el}}\Theta_\lambda^{\mathrm{nue}}$ are numerically coincident with corresponding elements $\langle \kappa_a'\lambda' \mid \mathsf{H}' \mid \kappa_a\lambda\rangle$ of the actual Hamiltonian; then the secular equations for the fictitious spin-only system will be identical, to first order, with those describing the actual system. Similar considerations allow us to define *second*-order terms in the spin Hamiltonian, such that $\langle M'\lambda' \mid \mathsf{H}_{\mathrm{S}}^{(2)} \mid M\lambda\rangle$ reproduces the second-order sum in the expression for $\langle \kappa_a'\lambda' \mid \mathsf{H}_{\mathrm{eff}} \mid \kappa_a\lambda\rangle$. There is no guarantee that we can find such operators with the properties

$$\langle M'\lambda' \mid \mathsf{H}_{\mathrm{S}}^{(1)} \mid M\lambda\rangle = \langle \kappa_a'\lambda' \mid \mathsf{H}' \mid \kappa_a\lambda\rangle \qquad \text{(first order)} \tag{5.5}$$

$$\langle M'\lambda' \mid \mathsf{H}_{\mathrm{S}}^{(2)} \mid M\lambda\rangle = \sum_{\kappa_b''\lambda''} \frac{\langle \kappa_a'\lambda' \mid \mathsf{H}' \mid \kappa_b''\lambda''\rangle\langle \kappa_b''\lambda'' \mid \mathsf{H}' \mid \kappa_a\lambda\rangle}{(E_a - E_{\kappa_b''})} \qquad \text{(second order)} \tag{5.6}$$

but in many cases we can, and then the use of a spin Hamiltonian is a perfectly valid procedure. We concentrate mainly on such cases but note, where appropriate, the factors involved in making further generalizations. The complete correspondence between the eigenvalue problems for the actual system and the pure spin system is summarized in Table 5.1 which puts the whole spin-Hamiltonian concept in a nutshell.

TABLE 5.1

CORRESPONDENCE BETWEEN ACTUAL SYSTEM AND SPIN-HAMILTONIAN "MODEL"[a]

	Actual system	Fictitious spin system
Eigenvalue equation	$\mathsf{H}\Psi = E\Psi$	$\mathsf{H}_S\Omega = E\Omega$
Expansion of wave function	$\Psi = \sum_{\kappa,\lambda} c_{\kappa\lambda}\Phi_\kappa\Theta_\lambda^{\text{nuc}}$	$\Omega = \sum_{M,\lambda} a'_{M\lambda}\Theta_M^{\text{el}}\Theta_\lambda^{\text{nuc}}$
Matrix eigenvalue equation	$\mathbf{Hc} = E\mathbf{c}$ (first form) $\mathbf{H}_{\text{eff}}\mathbf{a} = E\mathbf{a}$ (projected form)	 $\mathbf{H}_S\mathbf{a}' = E\mathbf{a}'$

[a] Note that Ψ is a function of all electronic variables and nuclear spins, while Ω is a function of electronic and nuclear spins only. The matrix eigenvalue equation in its first form is, in principle, infinite, involving a complete set of electronic functions Φ_κ ; the equivalent "projected" form contains only the coefficients (**a**) of the degenerate functions describing the states that are split by the perturbation; the corresponding equation for the fictitious spin-only system (right-hand column) takes a numerically identical form when the parameters in the spin Hamiltonian H_S are given suitable values. The necessary correspondence between the elements of $\mathbf{H}_{\text{eff}}$ and $\mathbf{H}_S$ is indicated in Eqs. (5.5) and (5.6).

Before considering specific effects arising from the many terms in H', we notice that each term gives an additive contribution to $\mathsf{H}_S^{(1)}$ (i.e., in first order), and each *pair* of terms gives an associated contribution to $\mathsf{H}_S^{(2)}$. Thus if we consider (introducing μ and ν as perturbation parameters for convenience in referring to orders of small quantities)

$$\mathsf{H}' = \mu\mathsf{H}_1 + \nu\mathsf{H}_2$$

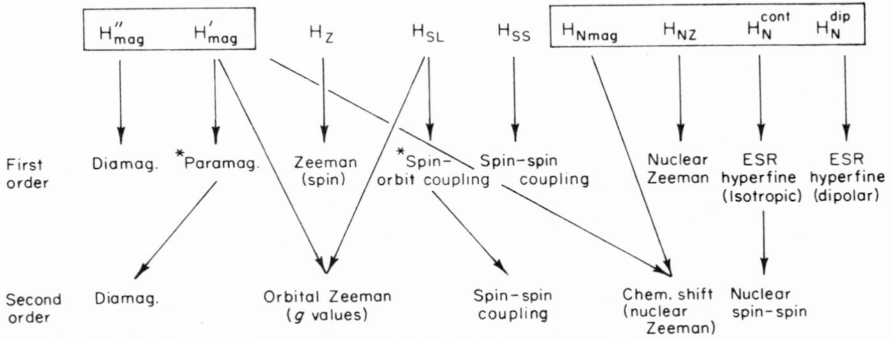

FIG. 5.1. Small terms in the Hamiltonian—some typical effects. Terms marked with an asterisk apply to free atoms (unquenched angular momentum). The paramagnetic term in this case gives also a first-order contribution to the Zeeman splitting.

the second-order terms will be of three types, a μ^2 term, a ν^2 term, and a $\mu\nu$ term. The "cross term" is often of special interest (as we saw in the simple example treated in Lecture 3); and it is useful at this stage to map out some of the territory to be explored. The main terms in H', as defined in (3.12)–(3.19), are set out on the top line in Fig. 5.1, and some of the first-order and second-order effects for which they are responsible are indicated below. Here we shall be considering only the first-order terms, but when we turn to NMR we shall find that even the simplest effects cannot be accounted for until we go to second order. For the moment, we shall consider one nuclear term, $\mathsf{H}_{\mathrm{N}}^{\mathrm{cont}}$, which is the "contact" term in the hyperfine Hamiltonian H_{N} of Eq. (3.19), and one purely electronic term, H_{SS} of Eq. (3.18), obtaining the first-order effects of each in turn.

Contact Hyperfine Coupling

Strong Field Case

The contact term in the full Hamiltonian is

$$\mathsf{H}_{\mathrm{N}}^{\mathrm{cont}} = (8\pi/3) \sum_{n,i} g\beta g_n\beta_{\mathrm{P}}\, \delta(\mathbf{r}_{ni})\, \mathbf{S}(i) \cdot \mathbf{I}(n) \tag{5.7}$$

and is responsible, as we shall see, for the spin-Hamiltonian term

$$\mathsf{H}_{\mathrm{S}}^{(1)}\,(\mathrm{cont}) = \sum_{n} a_n^{\mathrm{cont}}\mathbf{S} \cdot \mathbf{I}(n) \tag{5.8}$$

already used in our preliminary example (p. 92). Somehow, all the relevant information contained in the many-electron wave function is absorbed into the numerical coupling constants a_n^{cont}. This is the simplest possible case, and it is well known that a_n^{cont} depends on the *spin density* evaluated at nucleus n. We want to give a very general derivation of this result, valid for any kind of wave function we may care to use.

Let us first assume (remembering our first example) that the hyperfine splitting is small compared with the electronic Zeeman splitting, so that there is no appreciable mixing of different electronic states. In this case, which may be regarded as a "strong field" limit, the (contact) hyperfine structure of the various Zeeman levels will be obtained simply by averaging $\mathsf{H}_{\mathrm{N}}^{\mathrm{cont}}$ over the corresponding wave function, i.e., by taking the *diagonal* element. The energy shift is then given by

$$\delta E = \langle \kappa_a \lambda \mid \mathsf{H}_{\mathrm{N}}^{\mathrm{cont}} \mid \kappa_a \lambda \rangle$$

for each state $\Phi_{\kappa_a} \Theta_\lambda$ of the degenerate group. This may be written, in terms of the one-electron density matrix $\rho_1(\mathbf{x}_1; \mathbf{x}_1')$ as

$$\delta E = (8\pi/3)\, g\beta\beta_{\mathrm{p}} \sum_n g_n \int_{\mathbf{x}_1'=\mathbf{x}_1} \delta(\mathbf{r}_{n1})\, \mathbf{S}(1) \cdot \mathbf{I}(n)\, \rho_1(\mathbf{x}_1; \mathbf{x}_1')\, d\mathbf{x}_1 \tag{5.9}$$

But

$$\mathbf{S}(1) \cdot \mathbf{I}(n) = \mathsf{S}_x(1)\, \mathsf{I}_x(n) + \mathsf{S}_y(1)\, \mathsf{S}_y(n) + \mathsf{S}_z(1)\, \mathsf{I}_z(n)$$

and since ρ_1 contains only the spin factors $\alpha(s_1)\, \alpha^*(s_1')$ and $\beta(s_1)\, \beta^*(s_1')$, it is clear that only the z term can be effective [S_x, for example, will change the α into a β according to (2.9), and the result will vanish on integrating over spin]. The integral thus reduces to

$$\int_{\mathbf{r}_1'=\mathbf{r}_1} \mathsf{I}_z(n)\, \delta(\mathbf{r}_{n1}) \int_{s_1'=s_1} \mathsf{S}_z(1)\, \rho_1(\mathbf{x}_1; \mathbf{x}_1')\, ds_1\, d\mathbf{r}_1 = \mathsf{I}_z(n)\, Q_{\mathrm{S}}(\mathbf{R}_n)$$

since the spin integration yields the spin density function $Q_{\mathrm{S}}(\mathbf{r}_1; \mathbf{r}_1')$ for the state in question, as defined in (4.19), and the remaining integration simply puts $\mathbf{r}_1' = \mathbf{r}_1 = \mathbf{R}_n$ (position vector of nucleus n). Thus if we introduce the normalized spin density, with $Q_{\mathrm{S}} = MD_{\mathrm{S}}$ in accordance with (4.18), the expectation value (5.9) may be written

$$\delta E = \langle \kappa_a \lambda \mid \mathsf{H}_{\mathrm{N}}^{\mathrm{cont}} \mid \kappa_a \lambda \rangle = \langle M\lambda \mid \mathsf{H}_{\mathrm{S}}^{(1)}\,(\mathrm{cont}) \mid M\lambda \rangle \tag{5.10}$$

where

$$\mathsf{H}_{\mathrm{S}}^{(1)}\,(\mathrm{cont}) = (8\pi g\beta\beta_{\mathrm{p}}/3)\sum_n g_n D_{\mathrm{S}}(\mathbf{R}_n)\,\mathsf{S}_z\mathsf{I}_z(n) = \sum_n a_n^{\mathrm{cont}}\mathsf{S}_z\mathsf{I}_z(n) \tag{5.11}$$

the expectation value of S_z neatly putting in the factor M appropriate to whatever state is considered. The fact that we have obtained an $\mathsf{S}_z\mathsf{I}_z(n)$ interaction instead of the more general form $\mathbf{S}\cdot\mathbf{I}(n)$ assumed in (5.8) is due simply to neglect of the *mixing* between the electronic states, brought about by S_x and S_y terms which do, in fact, connect states differing in M values. Nevertheless, the derivation serves to show the value and generality of the concept of spin density and its connection with experimentally observable quantities.

Digression

The preceding result applies only in the strong field case because mixing of different product functions was neglected and it was accordingly sufficient to evaluate only diagonal matrix elements. More generally we must taken into account the *off*-diagonal terms $\langle\kappa_a'\lambda' \mid \mathsf{H}_{\mathrm{N}}^{\mathrm{cont}} \mid \kappa_a\lambda\rangle$ showing that these may be correctly reproduced as matrix elements $\langle M'\lambda' \mid \mathsf{H}_{\mathrm{S}}^{(1)}(\mathrm{cont})\mid M\lambda\rangle$ of the more complete operator $\sum_n a_n^{\mathrm{cont}}\mathbf{S}\cdot\mathbf{I}(n)$ used in (5.8) and then solving the corresponding spin-Hamiltonian secular problem. To do this we need to generalize slightly the notion of a density matrix, by admitting *transition* density matrices connecting different states, and we also need an important result due to Wigner and Eckart.

First we define the transition density matrices for $\Phi_\kappa \to \Phi_{\kappa'}$ by

$$\rho_1(\kappa\kappa' \mid \mathbf{x}_1; \mathbf{x}_1') = N\int \Phi_\kappa(\mathbf{x}_1, \mathbf{x}_2, \ldots, \mathbf{x}_N)\,\Phi_{\kappa'}^*(\mathbf{x}_1', \mathbf{x}_2, \ldots, \mathbf{x}_N)\,d\mathbf{x}_2 \cdots d\mathbf{x}_N \tag{5.12}$$

with a similar expression for $\rho_2(\kappa\kappa' \mid x_1, x_2; x_1', x_2')$. The quantity used in (5.9) is thus simply the special case $\kappa' = \kappa = \kappa_a$:

$\rho_1(\mathbf{x}_1; \mathbf{x}_1') = \rho_1(\kappa_a\kappa_a \mid \mathbf{x}_1; \mathbf{x}_1')$. Just as $\rho_1(\mathbf{x}_1; \mathbf{x}_1')$ determined all diagonal matrix elements of one-electron operators, so $\rho_1(\kappa\kappa' \mid \mathbf{x}_1; \mathbf{x}_1')$ determines the general (diagonal or off-diagonal) element:

$$\left\langle \kappa' \middle| \sum_i \mathsf{f}(i) \middle| \kappa \right\rangle = \int_{\mathbf{x}_1'=\mathbf{x}_1} \mathsf{f}(1)\, \rho_1(\kappa\kappa' \mid \mathbf{x}_1; \mathbf{x}_1')\, d\mathbf{x}_1 \tag{5.13}$$

and $\rho_2(\kappa\kappa' \mid \mathbf{x}_1, \mathbf{x}_2; \mathbf{x}_1', \mathbf{x}_2')$ has a similar property with respect to two-electron operators.

Next we recall that spin states have very characteristic transformation properties under a rotation of axis of quantization. If $\Theta_M^{(S)\prime}$ refers to an axis rotated with respect to some fixed frame, for which the corresponding spin eigenfunctions are $\Theta_M^{(S)}$, we find the $\Theta_M^{(S)\prime}$ are simply mixtures of the $\Theta_M^{(S)}$ and that a unique matrix is associated with each possible rotation; the matrices provide an "irreducible representation" of the three-dimensional rotation group, and each choice of S characterizes such a representation, with M labeling the different states "belonging to" that representation. Whenever we meet matrix elements connecting states of well-defined "symmetry species" (in this case labeled by S and M), we expect that reductions may be achieved by using group theory. The present case is no exception; basically we are interested in matrix elements connecting states Φ_κ and $\Phi_{\kappa'}$, of symmetry species (S, M) and (S', M') say, and even the operators $(\mathsf{S}_x, \mathsf{S}_y, \mathsf{S}_z)$ will also have well-defined transformation properties under a rotation of axes. To obtain maximum simplification, we work in terms of new combinations of the three spin operators, namely

$$\mathsf{S}_{+1} = -(\mathsf{S}_x + i\mathsf{S}_y)/\sqrt{2}, \qquad \mathsf{S}_0 = \mathsf{S}_z, \qquad \mathsf{S}_{-1} = (\mathsf{S}_x - i\mathsf{S}_y)/\sqrt{2} \tag{5.14}$$

because these have transformation properties *identical* with those of a set of spin states with quantum number $S = 1$. We can now state the key theorem on which most of spin-Hamiltonian theory depends:

If $\{\mathsf{T}_m^{(s)}\}$ is a set of $(2s+1)$ operators, behaving under rotations like a set of angular momentum functions with eigenvalues (s, m), while Φ_κ and $\Phi_{\kappa'}$ correspond similarly to eigenvalues (S, M) and (S', M'), respectively, then

$$\langle \kappa' \mid \mathsf{T}_m^{(s)} \mid \kappa \rangle = \text{constant} \times \begin{pmatrix} S & s & \vert & S' \\ M & m & \vert & M' \end{pmatrix} \tag{5.15}$$

where "constant" means a quantity independent of M, M', and m, and the other factor is a numerical coefficient determined purely by symmetry considerations; it is in fact the Clebsch–Gordan coefficient used in coupling states of systems with angular momenta S and s to one with a resultant S'. Such coefficients, which vanish unless $M + m = M'$, have been extensively tabulated.[1] Generally, the $\mathsf{T}_m^{(s)}$ are called "irreducible tensor operators of rank s," the "vector operators" S_0 and $\mathsf{S}_{\pm 1}$ corresponding to $s = 1$.

To see the connection with density matrices, we now consider the most general matrix element encountered in dealing with the hyperfine interaction. Discarding the nuclear factors, this is of the form

$$\left\langle \kappa' \left| \sum_i \mathsf{f}(i)\, \mathsf{S}_m(i) \right| \kappa \right\rangle$$

where $m = 0, \pm 1$, and $\mathsf{f}(i)$ is a *spatial* operator. Thus we may write, exactly as in the derivation of (4.17),

$$\left\langle \kappa' \left| \sum_i \mathsf{f}(i)\, \mathsf{S}_m(i) \right| \kappa \right\rangle = \int_{\mathbf{x}_1'=\mathbf{x}_1} \mathsf{f}(1)\, \mathsf{S}_m(1)\, \rho_1(\kappa\kappa' \mid \mathbf{x}_1; \mathbf{x}_1')\, d\mathbf{x}_1$$

$$= \int_{\mathbf{r}_1'=\mathbf{r}_1} \mathsf{f}(1)\, Q_S(\kappa\kappa' \mid \mathbf{r}_1; \mathbf{r}_1')_m\, d\mathbf{r}_1 \tag{5.16}$$

[1] A useful compilation appears in Condon and Shortley [7, pp. 76, 77], where the coefficient in (5.15) is denoted by $(SsMm \mid SsS'M')$. Various notations are in use, the one used here being chosen to resemble the Wigner 3-j symbol, to which it is simply related. For a full discussion of coupling coefficients and conventions, the reader is referred to the standard works on angular momentum (e.g., Edmonds [8]).

where $Q_S(\kappa\kappa' \mid \mathbf{r}_1; \mathbf{r}_1')_m$ is a *transition* spin density appropriate to states whose M values differ by m. Moreover, such matrix elements (and hence the corresponding spin densities) are related, for different values of M, M', and m, merely by a numerical factor; thus

$$Q_S(\kappa\kappa' \mid \mathbf{r}_1; \mathbf{r}_1')_m = \begin{bmatrix} S & 1 & | & S' \\ M & m & | & M' \end{bmatrix} Q_S(\bar{\kappa}\bar{\kappa}' \mid \mathbf{r}_1; \mathbf{r}_1') \tag{5.17}$$

where the square-bracket quantity is simply the ratio

$$\begin{bmatrix} S & 1 & | & S' \\ M & m & | & M' \end{bmatrix} = \begin{pmatrix} S & 1 & | & S' \\ M & m & | & M' \end{pmatrix} \Big/ \begin{pmatrix} S & 1 & | & S' \\ S & \bar{m} & | & S' \end{pmatrix} \qquad (\bar{m} = S' - S) \tag{5.18}$$

while $\bar{\kappa}$, $\bar{\kappa}'$ indicate two particular states, chosen here to be the "standard" states with $M = S$ and $M' = S'$, as used in Lecture 4. The "standard" density defined for $m = \bar{m} = S' - S$, is

$$Q_S(\bar{\kappa}\bar{\kappa}' \mid \mathbf{r}_1; \mathbf{r}_1') = \int_{s_1' = s_1} \mathsf{S}_{\bar{m}}(1)\, \rho_1(\bar{\kappa}\bar{\kappa}' \mid \mathbf{x}_1; \mathbf{x}_1')\, ds_1 \tag{5.19}$$

Evidently, for any other pair of states (M, M') there is only *one* nonzero density in (5.17), that for which $m = M' - M$.

Exactly similar considerations may be applied to the other coupling functions introduced in Lecture 4:

$$Q_{SL}(\kappa\kappa' \mid \mathbf{r}_1, \mathbf{r}_2; \mathbf{r}_1', \mathbf{r}_2')_m = \begin{bmatrix} S & 1 & | & S' \\ M & m & | & M' \end{bmatrix} Q_{SL}(\bar{\kappa}\bar{\kappa}' \mid \mathbf{r}_1, \mathbf{r}_2; \mathbf{r}_1', \mathbf{r}_2') \tag{5.20}$$

determines all matrix elements of $\sum'_{i,j} \mathsf{g}(i,j)\, \mathsf{S}_m(i)$, while

$$Q_{SS}(\kappa\kappa' \mid \mathbf{r}_1, \mathbf{r}_2; \mathbf{r}_1', \mathbf{r}_2')_m = \begin{bmatrix} S & 2 & | & S' \\ M & m & | & M' \end{bmatrix} Q_{SS}(\bar{\kappa}\bar{\kappa}' \mid \mathbf{r}_1, \mathbf{r}_2; \mathbf{r}_1', \mathbf{r}_2') \tag{5.21}$$

determines all matrix elements involving rank 2 spin operators. A rank 2 operator with $m = 0$ has already been used in defining Q_{SS}, namely

$$\mathsf{S}_0^{(2)}(i,j) = [2\mathsf{S}_z(i)\, \mathsf{S}_z(j) - \mathsf{S}_x(i)\, \mathsf{S}_x(j) - \mathsf{S}_y(i)\, \mathsf{S}_y(j)]$$

and the remainder of the set (with $m = \pm 1, \pm 2$) may also be constructed by linear combination of $\mathsf{S}_\mu(i)\,\mathsf{S}_\nu(j)$ $(\mu, \nu = x, y, z)$. The coefficient appearing in (5.21) is exactly analogous to that in (5.18) but with the 1 replaced by 2.

We are now in a position to complete the discussion of the contact hyperfine effects and, indeed, to deal with all the remaining spin terms in the Hamiltonian.

General Case

The most general matrix element we need is $\langle \kappa_a'\lambda' |\, \mathsf{H}_{\mathrm{N}}^{\mathrm{cont}}\, | \kappa_a \lambda \rangle$, where κ_a, κ_a' label degenerate states differing only in spin z component M. The spin scalar product may be expressed in terms of the S_m, I_m components of the type defined in (5.14), since $\mathbf{S}(i) \cdot \mathbf{I}(n) = \sum_m (-1)^m \mathsf{S}_m(i)\, \mathsf{I}_{-m}(n)$ (as is easily verified by expansion); we therefore consider a typical term of the form

$$\left\langle \kappa_a'\lambda' \left| \sum_i \delta(\mathbf{r}_{ni})\, \mathsf{I}_{-m}(n)\, \mathsf{S}_m(i) \right| \kappa_a \lambda \right\rangle$$

$$= \left\langle \kappa_a' \left| \sum_i \delta(\mathbf{r}_{ni})\, \mathsf{S}_m(i) \right| \kappa_a \right\rangle \langle \lambda' \,|\, \mathsf{I}_{-m}(n) \,|\, \lambda \rangle \qquad (5.22)$$

and note that the electronic factor reduces, using (5.16) and (5.17), to

$$\left\langle \kappa_a' \left| \sum_i \delta(\mathbf{r}_{ni})\, \mathsf{S}_m(i) \right| \kappa_a \right\rangle$$

$$= \begin{bmatrix} S & 1 & S' \\ M & m & M' \end{bmatrix} \int_{\mathbf{r}_1' = \mathbf{r}_1} \delta(\mathbf{r}_{n1})\, Q_{\mathrm{S}}(\bar{\kappa}_a \bar{\kappa}_a \,|\, \mathbf{r}_1; \mathbf{r}_1')\, d\mathbf{r}_1$$

$$= \begin{bmatrix} S & 1 & S' \\ M & m & M' \end{bmatrix} Q_{\mathrm{S}}(\bar{\kappa}_a \bar{\kappa}_a \,|\, \mathbf{R}_n)$$

The reduction to a spin Hamiltonian is immediate, in this case, because the coupling constant can be expressed alternatively in terms of matrix elements of a pure spin operator between states Θ_M. Thus, from (5.15),

$$\langle M' \mid \mathsf{S}_m \mid M \rangle = \begin{pmatrix} S & 1 & S \\ M & m & M' \end{pmatrix} \times \text{constant}$$

$$\langle S \mid \mathsf{S}_m \mid S \rangle = \begin{pmatrix} S & 1 & S \\ S & 0 & S \end{pmatrix} \times \text{constant}$$

and division gives (since $\langle S \mid \mathsf{S}_z \mid S \rangle = S$)

$$\begin{bmatrix} S & 1 & S \\ M & m & M' \end{bmatrix} = (1/S) \langle M' \mid \mathsf{S}_m \mid M \rangle \tag{5.23}$$

On introducing the normalized spin density $D_S(aa \mid \mathbf{r}_1; \mathbf{r}_1') = S^{-1} Q_S(\bar{\kappa}_a \bar{\kappa}_a \mid \mathbf{r}_1; \mathbf{r}_1')$, for the states of multiplet a, we find

$$\langle \kappa_a' \lambda' \mid \mathsf{H}_N^{\text{cont}} \mid \kappa_a \lambda \rangle = (8\pi g \beta \beta_P/3) \sum_n g_n D_S(aa \mid \mathbf{R}_n)$$
$$\times \sum_m (-1)^m \langle M' \mid \mathsf{S}_m \mid M \rangle \langle \lambda' \mid \mathsf{I}_{-m}(n) \mid \lambda \rangle$$

which may be written in the required form (5.5), namely

$$\langle \kappa_a' \lambda' \mid \mathsf{H}_N^{\text{cont}} \mid \kappa_a \lambda \rangle = \langle M' \lambda' \mid \mathsf{H}_S^{(1)}(\text{cont}) \mid M \lambda \rangle \tag{5.24}$$

where

$$\mathsf{H}_S^{(1)}(\text{cont}) = \sum_n a_n^{\text{cont}} \mathbf{S} \cdot \mathbf{I}(n) \tag{5.25}$$

and the coupling constant is the same as was defined in the strong-field case:

$$a_n^{\text{cont}} = (8\pi g \beta \beta_P/3) g_n D_S(aa \mid \mathbf{R}_n) \tag{5.26}$$

The contact hyperfine splitting is clearly isotropic (the coupling constants depending only on the spin density evaluated at *points*, and therefore being independent of molecular orientation).

The dipolar interaction term $\mathsf{H}_N^{\text{dip}}$ may be dealt with in a similar way, and gives a tensor interaction which is usually written in terms of Cartesian components, as indicated in an earlier Lecture (p. 80):

$$\mathsf{H}_{\mathrm{S}}^{(1)}\,(\mathrm{dip}) = \sum_{n}\sum_{\lambda,\mu} a_{n,\lambda\mu}^{\mathrm{dip}}\mathsf{S}_{\lambda}\mathsf{I}_{\mu}(n) \tag{5.27}$$

where, for instance,

$$\begin{aligned} a_{n,xx}^{\mathrm{dip}} &= (3g\beta g_n\beta_{\mathrm{P}}) \int r_{n1}^{-5}(x_{n1}^2 - \tfrac{1}{3}\,r_{n1}^2)\, D_{\mathrm{S}}(aa \mid \mathbf{r}_1)\, d\mathbf{r}_1 \\ a_{n,xy}^{\mathrm{dip}} &= (3g\beta g_n\beta_{\mathrm{P}}) \int r_{n1}^{-5}x_{n1}\,y_{n1} D_{\mathrm{S}}(aa \mid \mathbf{r}_1)\, d\mathbf{r}_1 \end{aligned} \tag{5.28}$$

($\mathbf{r}_{n1}$ being the vector from nucleus n to the point $\mathbf{r}_1$ at which the spin density is evaluated). The coupling constants therefore measure second moments of the spin density in the vicinity of each nucleus. Different experimental situations will therefore yield different kinds of information; for rapidly tumbling molecules, the tensor interactions average to zero, leaving only the isotropic coupling described by (5.26), but by using crystals it is possible to determine principal axes of the tensor interaction and thus to obtain information about the "shape" of the spin distribution near a nucleus.

Electron Spin-Spin Coupling

In fitting the ESR spectra of transition metal ions it is necessary to include a spin-Hamiltonian term of the form $\sum_{\mu,\nu} D_{\mu\nu}\mathsf{S}_{\mu}\mathsf{S}_{\nu}$; this term is appreciable when spin-orbit coupling is large and is now known to be a second-order consequence of the usual $\lambda\mathbf{L}\cdot\mathbf{S}$ interaction. In organic molecules and radicals, on the other hand, spin-orbit coupling is much smaller and second-order effects are quite negligible. Nevertheless it has been shown, first by Hutchison and Mangum [*9*] and then by van der Waals [*10*] and others, that an exactly similar spin-Hamiltonian term is necessary in order to explain the ESR spectra of molecules in triplet states in weak fields. It soon became apparent that the observed "zero-field splitting" of the Zeeman levels must be due to the direct spin-spin interactions contained in H_{SS} of (3.18). As in the case of H^{N}_{cont}, it is useful, in making a general theoretical analysis, to introduce tensor operators. The rank 2 operator we have already used, namely

$$[2\mathsf{S}_z(i)\,\mathsf{S}_z(j) - \mathsf{S}_x(i)\,\mathsf{S}_x(j) - \mathsf{S}_y(i)\,\mathsf{S}_y(j)]$$

is frequently denoted by $[\mathbf{S}(i) \times \mathbf{S}(j)]^{(2)}_0$ where the superscript (2) indicates the rank and the zero the particular component; the other four combinations of products are thus conveniently denoted by $[\mathbf{S}(i) \times \mathbf{S}(j)]^{(2)}_m$ with m taking the remaining values ± 1, ± 2. The spin-spin term may then be written[2]

$$\mathsf{H}_{SS} = -\tfrac{3}{2}\sum_{i,j}{}' g^2\beta^2 r_{ij}^{-5} \sum_m (-1)^m\, [\mathbf{r}_{ij} \times \mathbf{r}_{ij}]^{(2)}_{-m}\, [\mathbf{S}(i) \times \mathbf{S}(j)]^{(2)}_m \tag{5.29}$$

and a reduction parallel to that used in reducing the matrix elements of H^{N}_{cont} yields

$$\langle \kappa_a'\lambda' \mid \mathsf{H}_{SS} \mid \kappa_a\lambda\rangle = -\tfrac{3}{2} g^2\beta^2 \sum_m (-1)^m\, \delta_{\lambda'\lambda} \frac{\langle M' \mid [\mathbf{S} \times \mathbf{S}]^{(2)}_m \mid M\rangle}{\langle S \mid [\mathbf{S} \times \mathbf{S}]^{(2)}_0 \mid S\rangle}$$
$$\times \int r_{ij}^{-5}[\mathbf{r}_{ij} \times \mathbf{r}_{ij}]^{(2)}_{-m}\, \varrho_{SS}(\bar{\kappa}_a\bar{\kappa}_a \mid \mathbf{r}_1, \mathbf{r}_2)_0\, d\mathbf{r}_1\, d\mathbf{r}_2 \tag{5.30}$$

where the $[\mathbf{S} \times \mathbf{S}]^{(2)}_m$ are constructed from total spin operators, matrix elements being taken between pure spin states, and the spin-spin coupling function is defined essentially as in (4.32). This reduction again involves the elimination of the Clebsch–

[2] We discard the contact term in (3.18) because this gives only a small shift, common to all states of any given multiplet, and is therefore of no interest in the present application.

Gordan coefficients, as in the derivation of (5.23). The final result (5.30) is already in spin-Hamiltonian form and may be equated to

$$\left\langle M'\lambda' \left| \sum_m (-1)^m d_{-m}^{(2)} [\mathbf{S} \times \mathbf{S}]_m^{(2)} \right| M\lambda \right\rangle$$

where there are five independent coefficients $d_m^{(2)}$, each depending on an integration over the spin-spin coupling function Q_{SS} for the standard state ($M = S$) of the a multiplet. It is customary to rearrange this result in Cartesian form:

$$\mathsf{H}_S^{(1)}\,(\text{spin-spin}) = \sum_{\mu,\nu} d_{\mu\nu} \mathsf{S}_\mu \mathsf{S}_\nu \qquad (\mu, \nu = x, y, z) \tag{5.31}$$

where, e.g.,

$$\begin{aligned} d_{xx} &= 3g^2\beta^2 \int r_{12}^{-5}(x_{12}^2 - \tfrac{1}{3} r_{12}^2)\, D_{SS}(aa \mid \mathbf{r}_1, \mathbf{r}_2)\, d\mathbf{r}_1\, d\mathbf{r}_2 \\ d_{xy} &= 3g^2\beta^2 \int r_{12}^{-5} x_{12} y_{12} D_{SS}(aa \mid \mathbf{r}_1, \mathbf{r}_2)\, d\mathbf{r}_1\, d\mathbf{r}_2 \end{aligned} \tag{5.32}$$

and D_{SS} is the normalized function obtained on dividing $Q_{SS}(\bar{\kappa}_a \bar{\kappa}_a \mid \mathbf{r}_1, \mathbf{r}_2)$ by $S_a(2S_a - 1)$.

The results obtained above [*1*, *2*] are absolutely general and have been discussed also by McLachlan [*11*]. The forms of the wave functions have not been prescribed in any way; they may be MO functions, VB functions or, in principle, exact solutions of the nonrelativistic Schrödinger equation. Different approximations produce spin densities and spin-coupling functions of varying accuracy but these may always be compared in a meaningful way among themselves and against the results of direct observation. Many general theorems are implicit in these results; e.g., the fact that closed shells make no contributions to the hyperfine coupling constants is implicit in the form of D_S for a wave function of the "separated" form[3] $\mathsf{A}[\Phi_A \Phi_B]$,

[3] Antisymmetrized products describing any two weakly interacting groups of electrons (A and B) are discussed further in the appendix.

where Φ_A is a singlet function—the most general form possible in which one can speak of a closed shell (4). It also follows that there can be no spin-spin coupling in a doublet state, general theory [1] showing that D_{SS} must vanish identically unless $S > \frac{1}{2}$. These density functions, which reflect all those features of the wave function that bear on the coupling of spins, thus form the bridgehead between experiment and theory. If the first-order spin Hamiltonian gave an adequate representation of all the observed effects (remembering that there are indeed other first-order terms, such as nuclear quadrupole interactions, that we have not considered so far, all would be well. Unfortunately, the most innocent-looking effects—many of which are large enough to be quite easily observed (e.g., chemical shifts)—often arise only in second order; this means in effect that instead of averaging over *unperturbed* wave functions, we have to take into account the perturbation of the wave function. The problem of accounting for the corresponding spin-Hamiltonian parameters is then somewhat more difficult; we turn to this problem in the final lecture.

REFERENCES

1. McWeeny, R., and Mizuno, Y., *Proc. Roy. Soc. (London) Ser. A* **254**, 554 (1961).
2. McWeeny, R. *J. Chem. Phys.* **42**, 1717 (1965).
3. Pryce, M. H. L., *Proc. Phys. Soc. (London) Ser. A* **63**, 25 (1950).
4. Bleaney, B., and Stevens, K. W. H., *Rept. Progr. Phys.* **16**, 108 (1953).
5. Low, W., *Solid State Phys. Suppl.* **2**, (1960).
6. Griffith, J. S., "The Theory of Transition Metal Ions." Cambridge Univ. Press, London and New York, 1961.
7. Condon, E. U., and Shortley, G. H., "The Theory of Atomic Spectra." Oxford Univ. Press, London and New York, 1959.
8. Edmonds, A. R., "Angular Momentum in Quantum Mechanics." Princeton Univ. Press, Princeton, New Jersey, 1957.
9. Hutchison, C. A., and Mangum, B. W., *J. Chem. Phys.* **29**, 952 (1958).
10. van der Waals, J. H., and de Groot, M. S., *Mol. Phys.* **2**, 333 (1959).
11. McLachlan, A. D., *Mol. Phys.* **5**, 51 (1962).

6

THE g TENSOR, CHEMICAL SHIFTS, NUCLEAR SPIN-SPIN COUPLING

There is an embarassingly wide choice of possible "second-order" contributions to the spin Hamiltonian; fortunately many of these give effects which average to zero for rapidly tumbling molecules and are not important for ESR and NMR spectroscopy in the gas or liquid phase. In Fig. 5.1 we have indicated the origins of a few of the more important terms, including the isotropic nuclear spin-spin coupling (which originates in the electron-nuclear contact term in H_N but is transmitted via the electron distribution), the chemical shift (which is, in general, dependent upon molecular orientation but does not normally average to zero), and the g-tensor interaction (which is particularly important in crystal ESR, particularly for systems containing transition metal ions). Now we turn to the details.

THE g TENSOR

First let us consider the (spin) Zeeman term itself, which is simply $\mathsf{H}_Z = g\beta \sum_i \mathbf{B} \cdot \mathbf{S}(i)$. This is already in spin-Hamiltonian form since it contains no operators other than spin and thus, within the degenerate[1] group of states $\Phi_{\kappa_a}\Theta_\lambda$, it has matrix elements

$$\langle \kappa_a{}'\lambda' \mid \mathsf{H}_Z \mid \kappa_a\lambda \rangle = \langle M'\lambda' \mid \mathsf{H}_S^{(1)}(Z) \mid M\lambda \rangle \tag{6.1}$$

where

$$\mathsf{H}_S^{(1)}(Z) = g\beta\mathbf{B}\cdot\mathbf{S} \tag{6.2}$$

and $\mathbf{S}$ is the total spin. The phenomenological term, however, is

$$\mathsf{H}_S(Z) = \beta \sum_{\mu,\nu} g_{\mu\nu} B_\mu \mathsf{S}_\nu \tag{6.3}$$

where the elements of the g tensor may take values differing widely from those for a free electron. To obtain such a form, we must examine the second-order contributions to $\mathsf{H}_S(Z)$, which will arise from any two terms in H' that are linear in field components and electron spins, respectively. The obvious candidates are H'_{mag} and H_{SL}, and we know from the simple example considered in Lecture 3 that such terms, even in simplified form, are capable of giving the observed effects. We also note that, choosing real wave functions for the unperturbed states Φ_{κ_a}, both these operators have zero matrix elements between A-group functions, and therefore give no first-order effects.

We recall the form of the second-order terms (p. 95) and note that if we put

$$\mathsf{H}' = \mathsf{H}_1 + \mathsf{H}_2$$

with $\mathsf{H}_1 = \mathsf{H}_{SL}$ and $\mathsf{H}_2 = \mathsf{H}'_{mag}$, then the required "cross terms" arising from (5.6) are

$$\sum_{\kappa_b''\lambda''} (E_a - E_b)^{-1} [\langle \kappa_a{}'\lambda' \mid \mathsf{H}_1 \mid \kappa_b''\lambda'' \rangle \langle \kappa_b''\lambda'' \mid \mathsf{H}_2 \mid \kappa_a\lambda \rangle$$

$$+ \langle \kappa_a{}'\lambda' \mid \mathsf{H}_2 \mid \kappa_b''\lambda'' \rangle \langle \kappa_b''\lambda'' \mid \mathsf{H}_1 \mid \kappa_a\lambda \rangle] \tag{6.4}$$

[1] We continue to study the case of an *orbitally* nondegenerate set of states, so that the spin quantum number M_a is sufficient to label the different electronically degenerate members of the group.

where the summation now runs over all "excited" unperturbed states κ_b'' ($=b, S_b, M_b''$). We have put $E_{\kappa_b''} = E_b$ since b refers, in general, to a degenerate multiplet with κ_b'' labeling the individual states. The summation over κ_b'' thus amounts to summation over all M_b'' and over all excited multiplets b. Can we hope to find a spin Hamiltonian term $\mathsf{H}_S^{(2)}$ that can reproduce this expression in the form $\langle M'\lambda' \mid \mathsf{H}_S^{(2)} \mid M\lambda\rangle$ (where $M = M_a$ to reproduce the $(2S_a + 1)$-fold electronic degeneracy of the A group) and, if so, what will be the meaning of the parameters in terms of the electron distribution?

First, let us neglect the spin-other-orbit part of H_{SL}, namely the two-electron terms in (3.15); such terms are almost invariably discarded without question, although we shall include them later. The one-electron part of H_{SL} we write in the form

$$\mathsf{H}_1 = \mathsf{H}_{SL} = g\beta^2 \sum_{m,i} (-1)^m \mathsf{f}_{-m}(i)\, \mathsf{S}_m(i) \tag{6.5}$$

This is in the standard tensor form (cf. p. 104) with the spinless operator $\mathsf{f}_{-m}(i)$ behaving (for the spatial rotations) like the $-m$ component of a rank 1 tensor, and we immediately obtain [cf (5.22)]

$$\langle \kappa_a'\lambda' \mid \mathsf{H}_1 \mid \kappa_b''\lambda''\rangle = g\beta^2 \delta_{\lambda'\lambda''} \sum_m (-1)^m \left[\begin{matrix} S_b & 1 \\ M_b'' & m \end{matrix}\middle| \begin{matrix} S_a \\ M_a' \end{matrix}\right]$$

$$\times \int_{\mathbf{r}_1'=\mathbf{r}_1} \mathsf{f}_{-m}(1)\, Q_S(\bar{\kappa}_b\bar{\kappa}_a \mid \mathbf{r}_1; \mathbf{r}_1')\, d\mathbf{r}_1 \tag{6.6}$$

where the density function is between standard states with $M_b'' = S_b$, $M_a' = S_a$. In the general case, where $S_b \neq S_a$, it is not possible to express the square-bracket quantity in terms of spin matrix elements, as we did in deriving the first-order terms in the spin Hamiltonian.

Nevertheless, we continue with the H_2 term defined in (3.13); this is *spinless*, being

$$\mathsf{H}_2 = \mathsf{H}'_{\text{mag}} = \beta \sum_i \mathbf{B} \cdot \mathsf{L}(i) = \beta \sum_{m,i} (-1)^m B_{-m} \mathsf{L}_m(i) \tag{6.7}$$

and therefore

$$\langle \kappa_b''\lambda'' \mid \mathsf{H}_2 \mid \kappa_a\lambda \rangle = \beta\delta_{\lambda''\lambda}\delta_{M_b''M_a} \sum_m (-1)^m B_{-m} \times \int_{\mathbf{r}_1'=\mathbf{r}_1} \mathsf{L}_m(1)\, P_1(ab \mid \mathbf{r}_1; \mathbf{r}_1')\, d\mathbf{r}_1 \qquad (6.8)$$

Thus, H_2 connects the A group only with multiplets of the same spin multiplicity, and therefore we shall not *need* the Clebsch–Gordan coefficients (in the matrix elements of H_{SL}) for $S_b \neq S_a$. We can then use the usual trick (p. 105), putting $S_b = S_a = S$ and introducing $\langle M' \mid \mathsf{S}_m \mid M'' \rangle / S$ in place of the square bracket in (6.6). The first part of the second-order sum (6.4) can now be expressed as (summing over M_b'' in multiplet b)

$$g\beta^2 \sum_{\substack{b \\ (S_b=S_a)}} \frac{\delta_{\lambda'\lambda}\left[\sum_m (-1)^m \langle M' \mid \mathsf{S}_m \mid M \rangle\, Y_{-m}^{ba}\right]\left[\sum (-1)^{m'} B_{-m'} X_{m'}^{ab}\right]}{E_a - E_b} \qquad (6.9)$$

where

$$X_m^{ab} = \int_{\mathbf{r}_1'=\mathbf{r}_1} \mathsf{L}_m(1)\, P_1(ab \mid \mathbf{r}_1; \mathbf{r}_1')\, d\mathbf{r}_1 \qquad (6.10)$$

and, putting in the actual form of the operator f_{-m} as defined in (3.15),

$$Y_{-m}^{ba} = \sum_n Z_n \int_{\mathbf{r}_1'=\mathbf{r}_1} r_{n1}^{-3} \mathsf{M}_{-m}^n(1)\, D_S(ba \mid \mathbf{r}_1; \mathbf{r}_1')\, d\mathbf{r}_1 \qquad (6.11)$$

These are numerical quantities, determined by the transition charge and spin densities connecting multiplets a and b; and they behave (under rotation of axes) as vector components. They involve the angular momentum operators defined in (3.14) and (3.16).

If we transform to Cartesian components, and assume real unperturbed functions, the final result is

$$\mathsf{H}_S^{(2)}(\text{SL/mag}) = \beta \sum_{\mu,\nu} G_{\mu\nu} B_\mu \mathsf{S}_\nu \qquad (\mu, \nu = x, y, z) \qquad (6.12)$$

where

$$G_{\mu\nu} = 2g\beta^2 \sum_{b(\mathrm{S}_b=\mathrm{S}_a)} (E_a - E_b)^{-1} X_\mu^{ab} Y_\nu^{ba} \tag{6.13}$$

When this second-order contribution is added to the first-order (spin-only) term (6.2), we obtain the complete Zeeman term in the usual form

$$\mathrm{H_S}\,(\text{Zeeman}) = \beta \sum_{\mu,\nu} g_{\mu\nu} B_\mu \mathrm{S}_\nu \tag{6.14}$$

with

$$g_{\mu\nu} = g\delta_{\mu\nu} + G_{\mu\nu} \tag{6.15}$$

Again, this result is general and rigorous. Expressions (6.10) and (6.11) show that the values of the *g*-tensor components depend on the angular character of the transition spin densities, especially in the immediate vicinity of heavy nuclei, since the atomic number Z_n weights the contributions to Y_{-m}^{ba} in (6.11). If spin-other-orbit interactions are admitted [*1*], the only effect is to add a further term to Y_ν^{ba}, this term depending on the spin-orbit coupling function Q_{SL} connecting multiplets *a* and *b*. There appear to be good prospects of evaluating *g* tensors from some of the increasingly accurate molecular wave functions now becoming available.

Chemical Shifts

The positions of NMR peaks for any given nucleus depend markedly on its electronic environment, as was clear from Fig. 1.7; they may lie far away[2] from the position expected for a bare nucleus. These "chemical shifts" are well understood on a

[2] Typically by 10–100 parts per million; but the peaks are so well resolved in NMR spectroscopy that for practical purposes such differences give peaks that are "far apart."

qualitative level, but their theoretical prediction is a matter of considerable difficulty.

The observed shift is proportional to the applied field and can be described experimentally (in an isotropic situation) using a spin-Hamiltonian term $-\sum_n g_n \beta_p \mathbf{B}_n \cdot \mathsf{I}(n)$ in which $\mathbf{B}_n = (1 - \sigma_n)\mathbf{B}$ replaces the externally applied field $\mathbf{B}$. The parameter σ_n depends on the electronic environment of nucleus n and is usually referred to as a "shielding constant"; it recognizes the fact that an imposed field $\mathbf{B}$ sets up electron currents which produce a field opposing $\mathbf{B}$, these being responsible for molecular diamagnetism. Because of this connection with the theory of diamagnetism, it is interesting to obtain the chemical shift by a procedure somewhat different from that employed in dealing with the g tensor.

In order to get physical insight into the situation, it is convenient this time to start from an alternative treatment of the perturbation problem discussed in Lecture 3. Instead of writing down directly the second-order *energy* term for the perturbation

$$\mathsf{H}' = \mathsf{H}_1 + \mathsf{H}_2$$

we apply the perturbation in two stages, first H_1 and then H_2. It may then be shown that the chemical shift part of the second-order Hamiltonian, bilinear in H_1 and H_2 with the explicit form (6.4), can be rewritten as

$$\langle \kappa_a'\lambda' \mid \mathsf{H}_{\text{eff}} \mid \kappa_a\lambda \rangle_{\text{cross term}} = \langle \tilde{\kappa}_a'\lambda' \mid \mathsf{H}_2 \mid \tilde{\kappa}_a\lambda \rangle \qquad (6.16)$$

where $\tilde{\kappa}_a$ labels the *perturbed* state arising from κ_a under the perturbation H_1. Also, Φ_{κ_a} may be treated for this purpose as a nondegenerate state, provided H_1 has no matrix elements between κ_a and κ_a'. That the two expressions are equivalent is easily verified by using the usual formula for the first-order perturbation of the wave function.

The chemical shift is bilinear in applied field and nuclear spins, and the obvious terms to give a coupling via the electron distribution are thus [taking just one nucleus n, and considering the terms available in (3.13) and (3.19)]

$$\mathsf{H}_1 = \beta \sum_i \mathbf{B} \cdot \mathbf{L}(i) \tag{6.17}$$

$$\mathsf{H}_2 = 2\beta\beta_\mathrm{p} \sum_i g_n r_{ni}^{-3} \mathbf{I}(n) \cdot \mathbf{M}^n(i) \tag{6.18}$$

The first term by itself produces the diamagnetic currents in the electron distribution, while the second arises from the nuclear magnetic dipole interacting with electronic motion. Let us apply H_1 first and write the first-order correction up to terms linear in field components. The corresponding perturbed density functions may then be expressed in the form

$$P_1(\tilde{\kappa}_a\tilde{\kappa}_a \mid \mathbf{r}_1; \mathbf{r}_1') = P_1(\kappa_a\kappa_a \mid \mathbf{r}_1; \mathbf{r}_1') + \sum_\mu B_\mu f_\mu(\mathbf{r}_1; \mathbf{r}_1') \qquad (\mu = x, y, z) \tag{6.19}$$

(with similar expressions for other density functions). Such functions contain all information about the perturbed system, with its diamagnetic currents, to first order in the field.

We now turn to the nuclear dipole perturbation H_2, and evaluate the required second-order part of the effective Hamiltonian in the form (6.16); this has elements

$$\begin{aligned} \langle \kappa_a'\lambda' \mid \mathsf{H}_\mathrm{eff} \mid \kappa_a\lambda\rangle_\mathrm{chem\ shift} &= \langle \tilde{\kappa}_a'\lambda' \mid \mathsf{H}_2 \mid \tilde{\kappa}_a\lambda\rangle \\ &= 2\beta g_n\beta_\mathrm{P} \sum_m (-1)^m \langle \lambda' \mid \mathsf{I}_{-m}(n) \mid \lambda\rangle\, \delta_{\kappa_a'\kappa_a} \\ &\quad \times \int_{\mathbf{r}_1'=\mathbf{r}_1} r_{n1}^{-3} \mathsf{M}^n{}_m(1)\, P_1(\tilde{\kappa}_a\tilde{\kappa}_a \mid \mathbf{r}_1; \mathbf{r}_1')\, d\mathbf{r}_1 \end{aligned} \tag{6.20}$$

where the $\delta_{\kappa_a'\kappa_a}$ appears because spin-independent perturbations leave the spin classification of the states unchanged; and since the angular momentum operators $\mathsf{M}^n{}_m(1)$ are spinless, their matrix elements vanish unless κ_a' and κ_a have the same spin quantum numbers ($M_a' = M_a$). The right-hand side of (6.20), containing $\delta_{M_a'M_a}$ and $\langle\lambda' \mid \mathsf{l}_m(n) \mid \lambda\rangle$, may therefore be written as the matrix element of an operator independent of *electron* spin, namely

$$\langle\kappa_a'\lambda' \mid \mathsf{H}_{\text{eff}} \mid \kappa_a\lambda\rangle_{\text{chem shift}} = \langle M'\lambda' \mid g_n\beta_{\text{p}}\mathbf{I}(n) \cdot \mathbf{B}_n^{\text{ind}} \mid M\lambda\rangle \tag{6.21}$$

where $\mathbf{B}_n^{\text{ind}}$ plays the part of an induced field, evaluated at nucleus n, with components

$$B_{n,\mu}^{\text{ind}} = (e/mc) \int_{\mathbf{r}_1'=\mathbf{r}_1} r_{n1}^{-3}[\mathbf{r}_{n1} \times \boldsymbol{\pi}(1)]_\mu \, P_1(\tilde{\kappa}_a\tilde{\kappa}_a \mid \mathbf{r}_1; \mathbf{r}_1') \, d\mathbf{r}_1 \tag{6.22}$$

Now the perturbed density function has two parts, according to (6.19), one corresponding to the static charge distribution before application of the field and the other linear in field components, representing the field-proportional induced currents. Only the latter contributes to the integral, and we therefore obtain

$$B_{n,\mu}^{\text{ind}} = \sum_\nu B_\nu(e/mc) \int_{\mathbf{r}_1'=\mathbf{r}_1} r_{n1}^{-3}[\mathbf{r}_{n1} \times \boldsymbol{\pi}(1)]_\mu f_\nu(\mathbf{r}_1; \mathbf{r}_1') \, d\mathbf{r}_1 \qquad (\mu, \nu = x, y, z) \tag{6.23}$$

or, introducing a shielding *tensor* such that $\mathbf{B}_n^{\text{ind}} = \boldsymbol{\sigma}_n\mathbf{B}$,

$$B_{n,\mu}^{\text{ind}} = \sum_\nu \sigma_{n,\mu\nu} B_\nu = (\boldsymbol{\sigma}_n\mathbf{B})_\mu$$

The chemical shift terms in the effective Hamiltonian may thus be written as matrix elements of a spin-Hamiltonian term

$$\mathsf{H}_{\text{S}}^{(2)}\,(\text{chem shift}) = g_n\beta_{\text{p}}(\boldsymbol{\sigma}_n\mathbf{B}) \cdot \mathbf{I}(n) \tag{6.24}$$

in which the elements of the shielding tensor are

$$\sigma_{n,\mu\nu} = (e/mc) \int_{r_1'=r_1} r_{n1}^{-3}[\mathbf{r}_{n1} \times \boldsymbol{\pi}(1)]_\mu f_\nu(\mathbf{r}_1; \mathbf{r}_1') \, d\mathbf{r}_1 \tag{6.25}$$

and $f_\nu(\mathbf{r}_1; \mathbf{r}_1')$ is the perturbation of the density matrix (of the electronic ground state) due to unit field along the ν axis—a function describing the diamagnetically induced circulation about that axis.

If we combine the contribution (6.24) with the first-order nuclear Zeeman term, and allow for any number of nuclei, we obtain

$$\mathsf{H}_S\,(\text{nuc Zeeman}) = -\beta_P \sum_n g_n(\mathbf{1} - \boldsymbol{\sigma}_n)\,\mathbf{B}\cdot\mathbf{I}(n) \qquad (6.26)$$

where $(\mathbf{1} - \boldsymbol{\sigma}_n)$ in general stands for the tensor with components $(\delta_{\mu\nu} - \sigma_{n,\mu\nu})$.

In principle we may calculate the shielding tensor directly from the expression (6.25), but there are formidable computational difficulties in the accurate calculation of magnetic perturbations. What is more important, however, is that we now have a detailed understanding of the mechanism by which the chemical shift arises.

To complete the justification of a qualitative picture based on diamagnetic currents, we note that the expression (6.22) for $\mathbf{B}_n^{\text{ind}}$ has a purely classical interpretation. If we identify

$$J_\mu = m^{-1}[\pi_\mu P_1(\tilde{\kappa}_a\tilde{\kappa}_a \mid \mathbf{r}_1; \mathbf{r}_1')]_{\mathbf{r}_1'=\mathbf{r}_1} \qquad (6.27)$$

as the μ component of the current density (in electron units) at point $\mathbf{r}_1$, then

$$\mathbf{B}_n^{\text{ind}} = -\frac{1}{c}\int \frac{(\mathbf{r}_{n1} \times \mathbf{j})}{r_{n1}^3}\, d\mathbf{r}_1 \qquad (6.22a)$$

where $\mathbf{j} = -e\mathbf{J}$ is the current in electrostatic units. But this is simply the formula of Biot and Savart for the field due to an arbitrary distribution of currents. It follows without difficulty that J_μ, given in (6.27), is in fact the many-electron generalization of the standard quantum-mechanical probability current density for a system in the presence of a magnetic field. The current density expression given in most textbooks refers to the simple case of a one-particle system. For such a system, the density matrix reduces to the product $\psi(\mathbf{r})\,\psi^*(\mathbf{r}')$, and we easily obtain the familiar result (remembering that $\boldsymbol{\pi}$ contains the vector potential $\mathbf{A}$)

$$J_x = \frac{\hbar}{2mi}\left[\psi^*(\mathbf{r})\frac{\partial}{\partial x}\psi(\mathbf{r}) - \psi(\mathbf{r})\frac{\partial}{\partial x}\psi^*(\mathbf{r})\right] + \frac{e}{mc}A_x\psi(\mathbf{r})\,\psi^*(\mathbf{r})$$

with similar expressions for the y and z components. There is thus a rather complete justification for a pictorial interpretation of chemical-shift parameters in terms of currents induced in the electron distribution. This simple picture is often useful in accounting for the signs and relative magnitudes of shifts; thus, if there is an induced ring current (Fig. 6.1), the external field

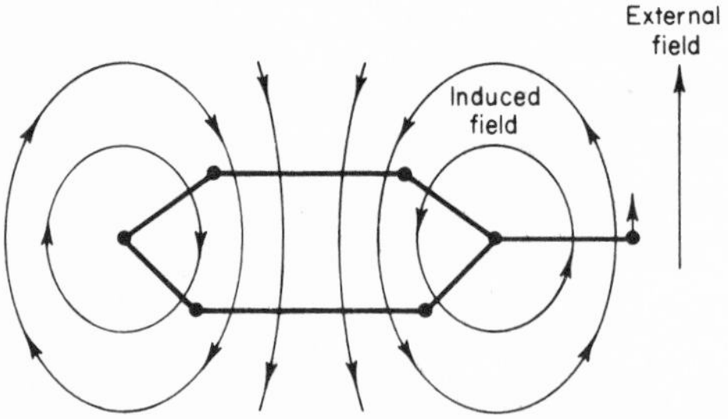

FIG. 6.1. Origin of the NMR chemical shift for a benzene proton. The external field induces a current around the benzene ring; at a peripheral proton, the external field is augmented by the corresponding induced field.

will be augmented at points outside the ring (e.g., at peripheral protons) but diminished at points inside the ring (and above or below). Such considerations can be particularly helpful in dealing with polycyclic molecules where the ring currents may be identified either intuitively or, more reliably, on the basis of simple quantum mechanical models. The predicted directions of various shifts are then often useful in making assignments of NMR signals to nuclei.

Nuclear Spin-Spin Coupling

As a final example, let us take the coupling between the spins of two nuclei. The main effect is the direct dipole-dipole interaction contained in the specifically nuclear part of the perturbation, H_N given in (4.19). This raises no problems because it is independent of electronic motion and is therefore already in spin-Hamiltonian form. The direct effect is of importance in dealing with crystals and leads to "broad line" NMR spectra from which important information about crystal parameters can be obtained. On the other hand, this dipole-dipole coupling averages to zero for rapidly tumbling molecules in solution, and the broad lines give way to exceedingly sharp lines. These are the lines of interest in high resolution NMR spectroscopy, and their origin was, for some time, a mystery. An explanation was given by Ramsey [*2*], who showed that these lines could arise from a variety of mechanisms involving an intermediate coupling with the electron distribution and that the most prominent effects (after averaging over all orientations) were those arising from the contact terms in H_{N}.

We shall confine our attention to two nuclei, n and n', and investigate the coupling due to their corresponding contact terms, taking

$$\begin{aligned} \mathsf{H}_1 &= (8\pi g\beta/3)\, g_n\beta_{\mathrm{p}} \sum_i \delta(\mathbf{r}_{ni})\, \mathbf{S}(i)\cdot\mathbf{I}(n), \\ \mathsf{H}_2 &= (8\pi g\beta/3)\, g_{n'}\beta_{\mathrm{p}} \sum_i \delta(\mathbf{r}_{n'i})\, \mathbf{S}(i)\cdot\mathbf{I}(n') \end{aligned} \tag{6.28}$$

The second-order sum (6.4) in the matrix of the effective Hamiltonian contains terms $\langle\kappa_a{}'\lambda' \mid \mathsf{H}_1 \mid \kappa_b''\lambda''\rangle$ and $\langle\kappa_b''\lambda'' \mid \mathsf{H}_2 \mid \kappa_a\lambda\rangle$, and similar terms with H_1 and H_2 reversed. We concentrate on the first sum, the second following by an appropriate interchange, and make the usual reduction [cf. (6.5) and (6.6)]

$$\langle \kappa_a{}'\lambda' \mid \mathsf{H}_1 \mid \kappa_b''\lambda'' \rangle = (8\pi g\beta/3)\, g_n \beta_{\mathrm{p}} \sum_m (-1)^m \langle \lambda' \mid \mathsf{I}_{-m}(n) \mid \lambda'' \rangle$$

$$\times \begin{bmatrix} S_b & 1 & S_a \\ M_b'' & m & M_a{}' \end{bmatrix} Q_{\mathrm{S}}(\bar{\kappa}_b \bar{\kappa}_a \mid \mathbf{R}_n) \qquad (6.29)$$

with a corresponding result for the other matrix element.

In the numerator of the second-order sum, namely $\langle \kappa_a{}'\lambda' \mid \mathsf{H}_1 \mid \kappa_b''\lambda'' \rangle \langle \kappa_b''\lambda'' \mid \mathsf{H}_2 \mid \kappa_a\lambda \rangle$ we now obtain a factor

$$\sum_{\lambda''} \langle \lambda' \mid \mathsf{I}_{-m}(n) \mid \lambda'' \rangle \langle \lambda'' \mid \mathsf{I}_{-m'}(n') \mid \lambda \rangle$$

and note that by the "closure property" (λ'' runs over a complete set of all possible nuclear spin products) this reduces simply to

$$\langle \lambda' \mid \mathsf{I}_{-m}(n)\, \mathsf{I}_{-m'}(n') \mid \lambda \rangle$$

When orbital factors are added, the sum therefore becomes

$$\sum_{\kappa_b''\lambda''} \frac{\langle \kappa_a{}'\lambda' \mid \mathsf{H}_1 \mid \kappa_b''\lambda'' \rangle \langle \kappa_b''\lambda'' \mid \mathsf{H}_2 \mid \kappa_a\lambda \rangle}{(E_a - E_b)}$$

$$= \sum_{m,m'} C_{mm'}(M_a{}', M_a) \langle \lambda' \mid \mathsf{I}_{-m}(n)\, \mathsf{I}_{-m'}(n') \mid \lambda \rangle \qquad (6.30)$$

where the coefficient $C_{mm'}(M_a{}', M_a)$ depends on $M_a{}'$ and M_a, but is not immediately of matrix element form. We do not yet know if it is possible to find an electron spin operator whose matrix elements will yield these values. The coefficient is, in fact, a complicated sum over all excited multiplet states κ_b''; as in dealing with the g tensor in (6.4) *et seq.*, we may write the coefficient as a sum over all levels b and all states ($M_b'' = S_b$,

$S_b - 1, \ldots, -S_b)$ within each level,

$$C_{mm'}(M_a', M_a) = (8\pi g\beta\beta_{\rm P}/3)^2 g_n g_{n'} \sum_b \sum_{M_b''} \frac{Q_{\rm S}(\bar{\kappa}_b\bar{\kappa}_a \mid \mathbf{R}_n) Q_{\rm S}(\bar{\kappa}_a\bar{\kappa}_b \mid \mathbf{R}_{n'})}{(E_a - E_b)}$$

$$\times (-1)^{m+m'} \begin{bmatrix} S_b & 1 & S_a \\ M_b'' & m & M_a \end{bmatrix} \begin{bmatrix} S_a & 1 & S_b \\ M_a & m' & M_b'' \end{bmatrix} \quad (6.31)$$

This is the first case in which we do not find a simple spin operator matrix element equivalent to the Clebsch–Gordan ratios in the square brackets. The crux of the difficulty is simply that we cannot set up an equivalence of the type (5.23) when $S_a \neq S_b$ because the matrix elements of pure spin operators S_m between spin states of different S always vanish. What we wish to show is that, for *all* the multiplets b that interact with multiplet a, the final sum in (6.31) can be written as the $M'M$ element of a single, suitably chosen electron spin operator. Thus, if $S_a = 1$, there would be multiplets with $S_b = 0, 1, 2$ interacting with multiplet a (these being the S_b values for which the Clebsch–Gordan coefficients are nonzero); and we should have to look at each case separately to find a suitable spin operator.

Fortunately, the case of most importance is that of a *singlet* electronic ground state (i.e., the only degeneracy then being due to *nuclear* spins), since the NMR spectra of molecules in paramagnetic ground states are exceedingly difficult to observe owing partly to the enormous effects of a free spin on the NMR signals, and partly to strong line broadening effects. For a singlet system, the analysis is simple and we very easily find a spin Hamiltonian of the expected form.

We put $S_a = 0$ in (6.31) and note that the only multiplets coupled with the ground state are then the triplets with $S_b = 1$ (all other Clebsch–Gordan coefficients vanishing). The coefficients also vanish unless $M_b'' + m = 0$ and $0 + m' = M_b''$, and the M_b'' summation therefore gives a factor

$$(-1)^{m+m'} \delta_{m,-m'} \begin{bmatrix} 1 & 1 & 0 \\ -m & m & 0 \end{bmatrix} \begin{bmatrix} 0 & 1 & 1 \\ 0 & -m & -m \end{bmatrix}$$

On putting in the values [$-(-1)^m$ and 1] of the coefficients, the coefficient $C_{mm'}$ is found to be

$$C_{mm'} = -\delta_{m,-m'}(-1)^m (8\pi g\beta\beta_{\text{p}}/3)^2 g_n g_{n'} Q_{\text{S}}(\bar{\kappa}_b\bar{\kappa}_a \mid \mathbf{R}_n)\, Q_{\text{S}}(\bar{\kappa}_a\bar{\kappa}_b \mid \mathbf{R}_{n'}) \tag{6.32}$$

This will be used in (6.30); and since

$$\sum_{m_1 m'} \delta_{m_1 -m'}(-1)^m \langle \lambda' \mid \mathsf{I}_{-m}(n)\, \mathsf{I}_{-m'}(n') \mid \lambda \rangle$$

$$= \left\langle \lambda' \,\middle|\, \sum_m (-1)^m\, \mathsf{I}_{-m}(n)\, \mathsf{I}_m(n') \,\middle|\, \lambda \right\rangle$$

the coupling term will finally involve a scalar product of nuclear spins. Thus, using (6.32) in (6.30) and adding a complex conjugate term, the second-order sum reduces to the matrix element of a spin-Hamiltonian term

$$\mathsf{H}_{\text{S}}^{(2)}\,(\text{nuc spin-spin}) = j_{nn'}\, \mathbf{I}(n) \cdot \mathbf{I}(n') \tag{6.33}$$

where (remembering that the two complex conjugate sums in (6.4) become identical for real wave functions) the coupling constant is

$$j_{nn'} = -2(8\pi g\beta\beta_{\text{p}}/3)^2 g_n g_{n'} \sum_{b(\text{triplet})} \frac{Q_{\text{S}}(\bar{\kappa}_a\bar{\kappa}_b \mid \mathbf{R}_{n'})\, Q_{\text{S}}(\bar{\kappa}_b\bar{\kappa}_a \mid \mathbf{R}_n)}{E_a - E_b} \tag{6.34}$$

There is one term of type (6.33) for each pair of nuclei, with a coupling constant determined by the formula (6.34); so once the transition spin densities connecting the ground state with the excited triplets are known, all the coupling constants follow at once. Again this result is rigorous and completely general, although the particular kind of many-electron wave function employed has not been stated; in other words, the same formula gives the coupling constants whether we use simple molecular orbital approximations, many-configuration functions, or even highly accurate functions involving interelectronic variables. Of course, if we specialize to the simplest possible representation of the ground and excited triplet states, using single or paired determinants of MO's and making various other approximations, the expression for $j_{nn'}$ reduces to the well-known forms given by McConnell [*3*], Pople and Santry [*4*], and others. If we adopt a VB approximation, it is most convenient to work from the general expression itself, using the VB expansions simply to obtain the transition spin density matrices. This approach has been followed by, e.g., Barfield [*5*] with considerable success, although at a somewhat more empirical level. Many-configuration wave functions have not yet been extensively used, but preliminary calculations of spin densities, etc., [*6*] show that there is no intrinsic difficulty in calculating more accurate density functions for inserting into the general expressions discussed in these last two lectures.

Conclusion

Perhaps it is now time to take stock of the current situation and of present theoretical ideas about spins in chemistry, and to look toward some more distant prospects in this area. We have followed the long route from relativistic quantum mechanics into the territory of chemical physics and have traced the origins of some of the main observable effects connected with electron and nuclear spins. With the development of improved methods of wave function calculation, the interpretation of bonding in terms of an apparent coupling of spins no longer occupies a prominent position in valence theory. With a nonrelativistic Hamiltonian the interactions are all "classical," and energy relationships can be understood qualitatively in terms of the charge density (P_1) and the pair function (P_2). At this level, spin makes itself felt only through the antisymmetry principle (e.g., by restricting the electron configurations available in an orbital description); its "dynamical" effects are confined to the "Fermi correlation" term in P_2, which ensures that $P_2^{\alpha\alpha}(\mathbf{r}_1, \mathbf{r}_2)$ and $P_2^{\beta\beta}(\mathbf{r}_1, \mathbf{r}_2) \rightarrow 0$ (like r_{12}^2) when two electrons of like spin approach. Other kinds of correlation, due to direct Coulomb repulsion, are notoriously difficult to allow for in wave-function calculations; but that which arises from spin is elegantly incorporated merely by antisymmetrizing the wave function. Many physical properties of the electron distribution however, such as its ability to scatter X rays (giving the crystallographer the possibility of "seeing" the molecule) depend purely on the form of the charge density P_1. This function, which is easily visualized as the electron density in a smeared-out charge cloud, therefore has an overwhelming importance. At this level then, spin has no *intrinsic* importance in chemistry; it could even be ignored altogether (as Matsen [7] and others have stressed) if we were prepared to accept equivalent *symmetry restrictions* on the (orbital) wave functions.

At the deeper level, however, when the relativistic terms are included in Pauli approximation (even though deep uncertainties remain), "real" spin interactions occur and lead into the vast and rapidly growing fields of experimental spectroscopy at microwave and radio frequencies—whose ramifications throughout chemistry, and more recently biology, we have not attempted to trace in these few lectures. Instead we have studied *concepts*: How are all these observable effects related to the basic spin interactions? The spin density (Q_S) is now almost as firmly established as the charge density in its role as the link between the many-electron wave function and the nuclear hyperfine effects observed in ESR. But we have seen that by introducing just *three* types of spin density function, Q_S, Q_{SL}, and Q_{SS}, a whole range of effects in both ESR and NMR—usually described through a phenomenological spin Hamiltonian—can be accounted for in completely general terms. Much work remains to be done in calculating these functions from wave functions of various types and varying accuracy; but this is largely a computational problem, quite separate from that of describing and interpreting the origin of spin-Hamiltonian parameters. The density functions are easy to visualize; they are related to the geometry of the molecule and describe, e.g., the amount of spin angular momentum in some region, or the dipolar coupling between electrons in specified volume elements; although they describe spin configurations, they are themselves *functions of spatial variables only*.

This brings us to the crux of the matter. Spin is a quantum concept, inseparable from a set of operator properties; whenever we try to "visualize" a molecule and discuss its behavior we fall back on classical models, because these represent the limitations of our everyday experience; and so, surprising as it may seem, the simplest way of dealing with spins in chemistry is to get the spins out of chemistry, or at least to separate them clearly from all our notions about the chemical bond. The wave function provides us with a few very fundamental spatial density functions. Two of these (P_1 and, to a lesser extent P_2) describe the chemical bonds and all spin-independent properties; the others provide an interpretation of the fine and hyperfine structure of molecular energy levels arising from tiny magnetic moments, which in themselves have no "chemical" significance but nevertheless enable us to obtain a wealth of information about the electronic structure from the empirically determined parameters in the spin Hamiltonian. Van Vleck's remark that the spin serves only as an *indicator* is still profoundly true; but who could have forseen, thirty years ago, just how useful an indicator it was going to become?

REFERENCES

1. McWeeny, R., *J. Chem. Phys.* **42**, 1717 (1965).
2. Ramsey, N. F., *Phys. Rev.* **91**, 303 (1953).
3. McConnell, H. M., *J. Chem. Phys.* **24**, 460 (1956).
4. Pople, J. A., and Santry, D. P., *Mol. Phys.* **8**, 1 (1964).
5. Barfield, M., *J. Chem. Phys.* **48**, 4458 (1968).
6. Cooper, I. L., and McWeeny, R., *J. Chem. Phys.* **49**, 3223 (1968).
7. Matsen, F. A., *Advan. Quantum Chem.* **1**, **60** (1964).

Appendix

THE INTERACTION OF TWO ELECTRONIC SYSTEMS[1]

In valence bond theory (Lecture 2) the interaction energy of two atoms, each with one valence electron, is related to the coupling of their spins, singlet coupling corresponding to bonding, triplet coupling corresponding to antibonding. We have noted, however, (pp. 42–44) that the possibility of relating the interaction energy between two systems to the coupling of their spins may be extended to systems of much greater complexity. In this appendix we discuss very generally the interaction energy of two electronic systems, let us call them A and B, and its representation (for systems in states with nonzero total spin) in terms of a formal spin coupling. We shall concentrate on the interaction energy itself because the total energy is of little direct importance; what matters is only the difference between the energies of the separate systems on the one hand, and the interacting systems on the other hand. In order to eliminate the total electronic energies of the separate subsystems, we shall assume their exact wave functions are (in principle) known and

[1] The material of this appendix did not form a part of the Science Development Lectures but was presented during a Quantum Chemistry Symposium held in the Chemistry Department, Polytechnic Institute of Brooklyn, on 22 March 1969. It is, however, so closely related to the substance of the lectures that it has been included in this volume.

may be used in building up the wave function of the composite system. This is reminiscent of the method of atoms in molecules [*1*], but the interacting systems are now molecules and the interactions are much weaker, suggesting that convergence will be very much better. We shall then use approximate wave functions in order to actually evaluate the interaction energy, for which we shall obtain an expression. We need consider basically only two distinct cases, though there are many possible applications. The two cases are

i. *A* and *B* in nondegenerate ground states (singlets). In this case there is no spin coupling problem, as such, and typical applications might be to the collisions of inert gas atoms or to the interactions which determine the packing of molecules in molecular crystals such as naphthalene.

ii. *A* and *B* in nonsinglet states. In this case there is a degeneracy whose resolution may be discussed in terms of spin coupling. This is the situation when two atoms, each with unpaired electrons, form a chemical bond; when two complete atomic shells interact to give a whole family of spectroscopic states; when two atoms or molecules react from excited triplet states; when systems in a crystal lattice are coupled to give ferromagnetism or antiferromagnetism; and in a great many other cases.

General Theory

We start in general from a set of product functions

$$\Psi_\kappa(\mathbf{x}_1, \mathbf{x}_2, \ldots, \mathbf{x}_N) = M_\kappa \mathsf{A}[\Phi_{Aa}(\mathbf{x}_1, \ldots, \mathbf{x}_{N_A}) \, \Phi_{Bb}(\mathbf{x}_{N_A+1}, \ldots, \mathbf{x}_{N_A+N_B})]$$

$$(\kappa = Aa, Bb) \qquad \text{(A.1)}$$

in which the label κ indicates a particular pair of factors, *Aa* and *Bb*, and *Aa*, for example, refers to system *A* in state *a*. The wave functions for the separate systems are assumed individually normalized and antisymmetrical and may in principle be exact molecular wave functions. A is the antisymmetrizer and M_κ is a normalizing factor. In the nondegenerate case (i) a single antisymmetrized product of this form may be a good description of the ground state of the whole system. More generally, however, we shall need to consider a wave function of the form

$$\Psi(\mathbf{x}_1, \mathbf{x}_2, ..., \mathbf{x}_N) = \sum_\kappa c_\kappa \Psi_\kappa(\mathbf{x}_1, \mathbf{x}_2, ..., \mathbf{x}_N) \tag{A.2}$$

where the coefficients are determined by solution of secular equations or, in the degenerate case, from symmetry considerations (e.g., in terms of coupling of the spins of the separate systems). If the wave functions of the two systems were strong-orthogonal [*2*] in the sense

$$\int \Phi^*_{Aa}(\mathbf{x}_1, \mathbf{x}_i, \mathbf{x}_j, ...)\, \Phi_{Bb}(\mathbf{x}_1, \mathbf{x}_k, \mathbf{x}_l, ...)\, d\mathbf{x}_1 = 0 \tag{A.3}$$

there would be no difficulty in evaluating the matrix elements contained in the energy expression. Unfortunately, however, we need to introduce the exact functions for *A* and *B* in order to be able to separate out the energies of the isolated systems, and in this case the functions will be to some extent nonorthogonal and (A.3) will not be satisfied. For interactions at long range where overlap is negligible the results become quite simple; but we are interested particularly in the much stronger interactions which occur at short and medium range, when the systems begin to interpenetrate appreciably. We must, therefore, study the general case of nonorthogonal interacting systems.

We recall that the necessary matrix elements can all be expressed in terms of a small number of density functions such as

$$\rho_1(\kappa\kappa' \mid \xi_1; \xi_1') = N \int \Psi_\kappa(\xi_1, \mathbf{x}_2, ..., \mathbf{x}_N)\, \Psi^*_{\kappa'}(\xi_1', \mathbf{x}_2, ..., \mathbf{x}_N)\, d\mathbf{x}_2 \cdots d\mathbf{x}_N \tag{A.4}$$

which is the one-electron transition matrix (5.12) connecting states Ψ_κ and $\Psi_{\kappa'}$. Here, for clarity in what follows, we have used ξ_1, ξ_1' for the variables appearing in the density matrix (i.e., referring to two points in configuration space rather than to the variables of a particular electron).

It is convenient to introduce an operator $\mathsf{O}_1(i)$, whose effect in any matrix element $\langle \Psi_{\kappa'} \mid \mathsf{O}_1 \mid \Psi_\kappa \rangle$ is to strike out the $\mathbf{x}_i$ integration, replacing $\mathbf{x}_i$ in Ψ_κ by ξ_1 and $\mathbf{x}_i$ in $\Psi_\kappa'^*$ by ξ_1'; and $\mathsf{O}_2(i)$ with a similar property except that ξ_1, ξ_1' are replaced by ξ_2, ξ_2'. On denoting the nonorthogonality integral formally by $\rho_0(\kappa\kappa')$ and adding a tilde to distinguish density matrices with respect to *non*orthogonal functions, we may then write [3]

$$\tilde{\rho}_0(\kappa\kappa') = \langle \Psi_{\kappa'} \mid \Psi_\kappa \rangle \tag{A.5a}$$

$$\tilde{\rho}_1(\kappa\kappa' \mid \xi_1; \xi_1') = \left\langle \Psi_{\kappa'} \middle| \sum_p \mathsf{O}_1(p) \middle| \Psi_\kappa \right\rangle \tag{A.5b}$$

$$\tilde{\rho}_2(\kappa\kappa' \mid \xi_1, \xi_2; \xi_1', \xi_2') = \left\langle \Psi_{\kappa'} \middle| {\sum_{p,q}}' \mathsf{O}_1(p)\, \mathsf{O}_2(q) \middle| \Psi_\kappa \right\rangle \tag{A.5c}$$

These functions completely determine all matrix elements [cf. p. (74)] of all one- and two-electron operators and thus lead to the energy expression corresponding to a variational wave function of the form (A.2). Thus, with the usual Hamiltonian (4.7), we obtain

$$\begin{aligned} H_{\kappa'\kappa} &= \langle \Psi_{\kappa'} \mid \mathsf{H} \mid \Psi_\kappa \rangle \\ &= \int_{\xi_1'=\xi_1} \mathsf{h}(1)\, \rho_1(\kappa\kappa' \mid \xi_1; \xi_1')\, d\xi_1 \\ &\quad + \tfrac{1}{2} \int_{\substack{\xi_1'=\xi_1 \\ \xi_2'=\xi_2}} g(1,2)\, \rho_2(\kappa\kappa' \mid \xi_1, \xi_2; \xi_1', \xi_2')\, d\xi_1\, d\xi_2 \end{aligned} \tag{A.6}$$

and

$$E = \sum_{\kappa,\kappa'} c_\kappa c_{\kappa'}^* H_{\kappa'\kappa} \tag{A.7}$$

In the same way, the density matrices in (A.5) determine all other electronic properties of the composite system AB. If we want to express the properties of AB in terms of those of the separate subsystems A and B, the basic problem is thus to express $\tilde{\rho}_0$, $\tilde{\rho}_1$ and $\tilde{\rho}_2$ in terms of ρ_n^A and ρ_m^B. The density matrices for normalized functions are of course simply $\tilde{\rho}_n = \tilde{\rho}_n/\tilde{\rho}_0$.

To reduce the density matrices, we write the antisymmetrizer in (A.1) as

$$\mathsf{A} = (N!)^{-1} \sum_P (-1)^p P \tag{A.8}$$

where the "normalization" ensures that $\mathsf{A}^2 = \mathsf{A}$ (i.e., the operator is "idempotent"), and use the well-known result

$$\mathsf{A} = \frac{N_A! N_B!}{(N_A + N_B)!} \mathsf{A}' \mathsf{A}_A \mathsf{A}_B \tag{A.9}$$

where A_A and A_B refer to the individual systems A and B, while A' is a sum involving all possible transpositions of variables *between* the two systems. It is then convenient to relabel the B-system variables $\mathbf{x}_{\bar{1}}$, $\mathbf{x}_{\bar{2}}$,... instead of $\mathbf{x}_{N_A+1}$, $\mathbf{x}_{N_A+2}$,... (so that $\bar{r} = N_A + r$) and to write $(i, \bar{r})$ for the operator which interchanges variables $\mathbf{x}_i$ (system A) and $\mathbf{x}_{\bar{r}}$ (system B). With this notation

$$\mathsf{A}' = \sum_{n=0}^{N_{\min}} (-1)^n \mathsf{P}_n \tag{A.10}$$

where P_n is a sum of multiple transpositions of the form

$$\mathsf{P}_n = \sum_{i_1 \cdots i_n} \sum_{\bar{r}_1 \cdots \bar{r}_n} (i_1, \bar{r}_1) \cdots (i_n, \bar{r}_n) \tag{A.11}$$

and $N_{\min}$ is the smaller N_A, N_B. For $n = 0$ we have, of course, the "zero transposition" which leaves the product unchanged and is thus equivalent simply to multiplication by 1.

We now assume that Φ_{Aa} and Φ_{Bb} are already normalized antisymmetric wave functions, so that A_A and A_B may be dropped. It follows easily that

$$\langle \Psi_\kappa \mid \Psi_\kappa \rangle = M_\kappa^2 (N_A! N_B!/N!) \langle \Phi_{Aa} \Phi_{Bb} \mid \mathsf{A}' \mid \Phi_{Aa} \Phi_{Bb} \rangle$$

and it is convenient to choose $M_\kappa = (N_A! N_B!/N!)^{-1/2}$ so as to ensure that Ψ_κ is normalized for large separation of the systems (where A' is equivalent to the unit operator).

With this normalization it follows readily that

$$\tilde{\rho}_0(\kappa\kappa') = \langle \Phi_{Aa'} \Phi_{Bb'} \mid \mathsf{A}' \mid \Phi_{Aa} \Phi_{Bb} \rangle$$

$$\tilde{\rho}_1(\kappa\kappa' \mid \xi_1; \xi_1') = \left\langle \Phi_{Aa'} \Phi_{Bb'} \middle| \sum_p \mathsf{O}_1(p)\, \mathsf{A}' \middle| \Phi_{Aa} \Phi_{Bb} \right\rangle \tag{A.12}$$

$$\tilde{\rho}_2(\kappa\kappa' \mid \xi_1, \xi_2; \xi_1', \xi_2') = \left\langle \Phi_{Aa'} \Phi_{Bb'} \middle| \sum_{p,q}{}' \mathsf{O}_1(p)\, \mathsf{O}_2(q)\, \mathsf{A}' \middle| \Phi_{Aa} \Phi_{Bb} \right\rangle$$

and that for large separation (or with strong-orthogonal functions) the tildes may be discarded. The results of reducing these expressions have been given elsewhere [*4*], and we therefore consider only one example. If we take the single-interchange part of A' in the first equation of (A.12) we obtain

$$\left\langle \Phi_{Aa'} \Phi_{Bb'} \middle| \sum_{i,\bar{r}} (i, \bar{r}) \middle| \Phi_{Aa} \Phi_{Bb} \right\rangle$$

$$= N_A N_B \langle \Phi_{Aa'} \Phi_{Bb'} \mid (1, \bar{1}) \mid \Phi_{Aa} \Phi_{Bb} \rangle$$

$$= N_A N_B \int \Phi^*_{Aa'}(\mathbf{x}_1, \mathbf{x}_2, \ldots, \mathbf{x}_{N_A})\, \Phi^*_{Bb'}(\mathbf{x}_{\bar{1}}, \mathbf{x}_{\bar{2}}, \ldots, \mathbf{x}_{\bar{N}_B})$$

$$\times\, \Phi_{Aa}(\mathbf{x}_{\bar{1}}, \mathbf{x}_2, \ldots, \mathbf{x}_{N_A})\, \Phi_{Bb}(\mathbf{x}_1, \mathbf{x}_{\bar{2}}, \ldots, \mathbf{x}_{\bar{N}_B})$$

$$\times\, d\mathbf{x}_1 \cdots d\mathbf{x}_{N_A}\, d\mathbf{x}_{\bar{1}} \cdots d\mathbf{x}_{\bar{N}_B}$$

But by definition of the one-electron density matrix (cf. A.4) this amounts to

$$\left\langle \Phi_{Aa'} \Phi_{Bb'} \middle| \sum_{i,\bar{r}} (i, \bar{r}) \middle| \Phi_{Aa} \Phi_{Bb} \right\rangle = \int \rho_1^A(\mathbf{x}_{\bar{1}}; \mathbf{x}_1)\, \rho_1^B(\mathbf{x}_1; \mathbf{x}_{\bar{1}})\, d\mathbf{x}_1\, d\mathbf{x}_{\bar{1}}$$

Multiple-interchange terms may be evaluated similarly and the final result is thus that $\tilde{\rho}_0(\kappa\kappa')$ is given by

$$1 - \int \rho_1^A(\mathbf{x}_1; \mathbf{x}_1') \, \rho_1^B(\mathbf{x}_1'; \mathbf{x}_1) \, d\mathbf{x}_1 \, d\mathbf{x}_1'$$
$$+ (1/2!) \int \rho_2^A(\mathbf{x}_1, \mathbf{x}_2; \mathbf{x}_1', \mathbf{x}_2') \, \rho_2^B(\mathbf{x}_1', \mathbf{x}_2'; \mathbf{x}_1, \mathbf{x}_2)$$
$$\times \, d\mathbf{x}_1 \, d\mathbf{x}_1' \, d\mathbf{x}_2 \, d\mathbf{x}_2'$$
$$- \cdots \tag{A.13}$$

where the names of the (dummy) variables have been changed merely for typographical convenience. In this way, the density matrices of the composite system AB may be expressed in terms of the density matrices of its parts. Further reductions (e.g., to the spinless densities) may be effected, as we shall see, in any particular case. These basic expansions, all of them finite with only $N_{\min}$ terms, have been written out in full elsewhere [*4*].

Applications

We now consider in more detail the two main types of application:

i. The nondegenerate systems, fairly well represented by a single term of (A.2).

ii. The degenerate case, in which the spins S_A and S_B must be vector-coupled to a resultant S by putting together several terms with symmetry-determined coefficients.

In the context of these lectures, we are more concerned with (ii) and with showing that the interaction can be described formally in terms of an effective spin Hamiltonian; but for completeness we include (i) in order to emphasize the connection with the theory of intermolecular forces—as usually developed for systems in singlet states and in the long-range region.

Nondegenerate Case

This case applies to systems in singlet states, $S_A = S_B = 0$. We consider first long-range interactions, for which non-orthogonality may be neglected and the density matrix expansions reduce to their first terms.

Long-Range Interactions[2]

The one-electron and two-electron density matrices for a one-term approximation $\Psi = \Psi_\kappa$ reduce as follows

$$\rho_1(\mathbf{x}_1; \mathbf{x}_1') = \rho_1^A(aa \mid \mathbf{x}_1; \mathbf{x}_1') + \rho_1^B(bb \mid \mathbf{x}_1; \mathbf{x}_1')$$

$$\begin{aligned}\rho_2(\mathbf{x}_1, \mathbf{x}_2; \mathbf{x}_1', \mathbf{x}_2') = {} & \rho_2^A(aa \mid \mathbf{x}_1, \mathbf{x}_2; \mathbf{x}_1', \mathbf{x}_2') + \rho_2^B(bb \mid \mathbf{x}_1, \mathbf{x}_2; \mathbf{x}_1', \mathbf{x}_2') \\ & + \rho_1^A(aa \mid \mathbf{x}_1; \mathbf{x}_1')\, \rho_1^B(bb \mid \mathbf{x}_2; \mathbf{x}_2') \\ & - \rho_1^A(aa \mid \mathbf{x}_2; \mathbf{x}_1')\, \rho_1^B(bb \mid \mathbf{x}_1; \mathbf{x}_2') \\ & + \rho_1^A(aa \mid \mathbf{x}_2; \mathbf{x}_2')\, \rho_1^B(bb \mid \mathbf{x}_1; \mathbf{x}_1') \\ & - \rho_1^A(aa \mid \mathbf{x}_1; \mathbf{x}_2')\, \rho_1^B(bb \mid \mathbf{x}_2; \mathbf{x}_1')\end{aligned} \qquad (A.14)$$

The first result (on removing the primes and integrating over spins) simply shows that, in this approximation, the electron density is the sum of the densities of system A (in state a) and B (in state b); the second shows that when A and B are well localized, and $\mathbf{x}_1$ and $\mathbf{x}_2$ are points in A and B, respectively, then the pair function reduces to the product form $\rho_1^A(\mathbf{x}_1)\ \rho_1^B(\mathbf{x}_2)$, showing that correlation is neglected at large distances. Thus this approximation is not sufficient to account for van der Waals' forces, which depend on correlation of electronic motions and require the addition of further terms in the expansion (A.2), as will be clear presently.

The one-term energy expression is

$$E = H^A + H^B + J^{AB} - K^{AB} \qquad (A.15)$$

[2] This case is discussed more fully elsewhere [5].

where, for instance,

$$H^A = \langle \Phi_{Aa} \mid \mathsf{H}^A \mid \Phi_{Aa} \rangle \tag{A.16}$$

represents the energy of the N_A electrons of group A, described by wave function Φ_{Aa}, alone in the field of the nuclei (of *both* groups), and[3]

$$\begin{aligned} J^{AB} &= \int \frac{\rho_1^A(aa \mid \mathbf{x}_1)\, \rho_1^B(bb \mid \mathbf{x}_2)}{r_{12}}\, d\mathbf{x}_1\, d\mathbf{x}_2 \\ K^{AB} &= \int \frac{\rho_1^A(aa \mid \mathbf{x}_2; \mathbf{x}_1)\, \rho_1^B(bb \mid \mathbf{x}_1; \mathbf{x}_2)}{r_{12}}\, d\mathbf{x}_1\, d\mathbf{x}_2 \end{aligned} \tag{A.17}$$

Thus J^{AB} represents simply the Coulomb interaction of the electronic charge densities of systems A and B, while K^{AB} is a generalization of the "exchange" term encountered in Lecture 2 and is very small for well-localized (nonoverlapping) systems.

The Hamiltonian in (A.16) is

$$\mathsf{H}^A = \sum_{i=1}^{N_A} [-\tfrac{1}{2}\nabla^2(i) + V_A(i) + V_B(i)] + \tfrac{1}{2} \sum_{i,j=1}^{N_A}{}' g(i,j) \tag{A.18}$$

where $V_A(i)$ is the potential energy of electron i in the field of the nuclei of system A, and refers only to N_A electrons. Simple rearrangement of (A.15) then yields

$$\begin{aligned} E = E^A + E^B &+ \int V_B(1)\, \rho_1^A(aa \mid \mathbf{x}_1)\, d\mathbf{x}_1 \\ &+ \int V_A(1)\, \rho_1^B(bb \mid \mathbf{x}_1)\, d\mathbf{x}_1 + J^{AB} - K^{AB} \end{aligned} \tag{A.19}$$

Here E^A and E^B are the energies of the two *separate* systems, A and B; the next two terms give the attractions between the

[3] For diagonal elements, we use the usual notation (4.10), e.g., $\rho_1^A(aa \mid \mathbf{x}_1; \mathbf{x}_1) = \rho_1^A(aa \mid \mathbf{x}_1)$.

nuclei of one system and the charge cloud of the other, while the remaining terms represent the (Coulomb-exchange) repulsion of the two charge clouds. The result has a purely classical significance and the energy of interaction is clearly separated from the (in principle *exact*) energies of the individual systems.

Usually, owing to the electrical neutrality of the interacting molecules, the electrostatic interactions are slight, except for molecules with strongly polar bonds. In this case the one-term approximation does not adequately represent the interactions, which arise largely from *correlation* of electronic motions in the two systems, as first recognized by London [6]. To discuss these effects, we simply add the higher terms in the expansion (A.2). These are of two types: those in which *one* system (A or B) is in an excited state, and those in which *both* systems are excited. A simple application of perturbation theory then shows that the single-excitation and double-excitation contributions are (to second order in the energy)

$$E^{(1)} = -\sum_{a'(\neq a)} \frac{|H^A_{\text{eff}}(aa')|^2}{E(a \to a')} - \sum_{b'(\neq b)} \frac{|H^B_{\text{eff}}(bb')|^2}{E(b \to b')} \tag{A.20}$$

and

$$E^{(2)} = -\sum_{\substack{a'(\neq a)\\ b'(\neq b)}} \frac{|J^{AB}(aa', bb') - K^{AB}(aa', bb')|^2}{E(a \to a', b \to b')} \tag{A.21}$$

Here, for instance,

$$H^A_{\text{eff}}(aa') = H^A(aa') + J^{AB}(aa', bb) - K^{AB}(aa', bb)$$

and the constituent terms are obvious generalizations of those introduced in (A.16) and (A.17). A full discussion of the energy contributions [5] shows that (A.20) arises from the *polarization* of one molecule by the field of the other; for neutral molecules without strongly polar bonds, this term is small, like the direct electrostatic interactions. The "dispersion" interactions, recognized by London, are contained in (A.21); at long range the K^{AB} term is negligible, while J^{AB} takes the form

$$J^{AB}(aa', bb') = \int \frac{\rho_1^A(aa' \mid \mathbf{x}_1)\, \rho_1^B(bb' \mid \mathbf{x}_2)}{r_{12}}\, d\mathbf{x}_1\, d\mathbf{x}_2 \qquad \text{(A.22)}$$

Spin integrations may be performed at once, the ρ's then being replaced by P's (i.e., charge densities) as in Lecture 4, and (A.22) thus represents the electrostatic interaction of the *transition* densities corresponding to virtual excitations $a \rightarrow a'$ in system A and $b \rightarrow b'$ in system B. The summation in (A.21) is over all such excitations, each pair making a contribution to an overall attraction between the molecules. It should be noted, as appears to have been pointed out first by Longuet–Higgins [7], that there is no need to follow the conventional procedure of making a multipole expansion of the interactions. In fact, for large molecules not too far apart, the multipole expansion breaks down and the interaction must be described in terms of the transition charge densities themselves.

After discussing the origin of the weak long-range attractions (experimentally recognized as the van der Waals' forces), we shall want to know the origin of the repulsions that occur at shorter range and keep the molecules apart. These short-range interactions rapidly become very large, by comparison with the dispersion interactions, as soon as the two systems begin to "interpenetrate" even slightly. To study this effect we may therefore revert to a one-term approximation.

Short-Range and Medium-Range Interactions

For molecules in nondegenerate singlet states, the density matrix expansions such as (A.13) may be reduced easily by integration over spins to give spinless quantities $\tilde{P}_0, \tilde{P}_1, \tilde{P}_2$. Thus

$$\tilde{P}_\nu = \sum_{n=1}^{N_{\min}} (-1)^n \tilde{P}_\nu^{(n)} \qquad (\nu = 0, 1, 2) \tag{A.23}$$

where $\tilde{P}_\nu^{(n)}$ arises from the n-interchange term in (A.10). The single interchange term introduces the major nonorthogonality corrections; we shall not discuss their mathematical forms at this stage, but their effects may be summarized in a few words. The interesting thing about these corrections is that they directly affect the *electron density* in the system; this is no longer just a superposition of densities for A and B separately; the corrections contain terms corresponding to charge being "pushed out" of the overlap region. The energy contributions arising from this effect (associated with the energy of the electron distribution in the field of the nuclei) are strongly repulsive, and the results of actual calculations have been fully discussed [*8*] for the case of two hydrogen molecules.

Repulsions due to modifications of the electron density between two systems have been encountered before (Lecture 2) in the triplet state of the hydrogen molecule—the "exchange" part of the energy arising in that case largely from electron density being diminished in the bond region, whereas for the singlet coupling it was augmented, giving molecular binding. This gives us an important clue to the probable behavior of interacting molecules in *non*singlet states. The resultant spins, S_A and S_B say, may then be coupled parallel or antiparallel (or to an intermediate resultant spin), and we suspect that the strong interactions acompanying interpenetration of charge clouds will then occur with either sign—just as in the Heitler–London calculation—parallel coupling favoring repulsion, antiparallel coupling leading to strong attractions. It is this spin-dependent interaction that is particularly relevant to the theme of the lectures, since once again, it leads to interactions that may be simulated by means of $\mathbf{S}^A \cdot \mathbf{S}^B$ terms in a spin Hamiltonian, as was anticipated in Lecture 2 (p. 43). We therefore pass directly to the general analysis, which includes $S_A = S_B = 0$ as a special case.

DEGENERATE CASE

For systems with spins S_A and S_B, there are $(2S_A + 1)(2S_B + 1)$ functions of the type (A.1) that are degenerate in the absence of interaction, the state labels a and b now being equivalent to spin quantum numbers M_A and M_B labeling the multiplet components. This degeneracy is lifted by interaction but, with a spinless Hamiltonian, the total spin is still described by good quantum numbers (S, M say), and the different values of S label the different branches of the energy as a function of separation. The vector-coupled product functions are

$$\Phi_{SM} = \sum_{M_A, M_B} \Phi_{A,M_A} \Phi_{B,M_B} \begin{pmatrix} S_A & S_B & | & S \\ M_A & M_B & | & M \end{pmatrix} \tag{A.24}$$

each of which yields a satisfactory wave function on antisymmetrizing:

$$\Psi_{SM} = M_{SM} \mathsf{A} \Phi_{SM} \tag{A.25}$$

Since the expansion coefficients are determined in this way by spin requirements, there is no need to solve a secular problem; we merely compute the energy expectation value, using (A.25), or the density matrices if more general properties are required. The spinless densities in this case take the form

$$\tilde{P}_\nu = \sum_{M_A, M_B} \sum_{M_A', M_B'} \begin{pmatrix} S_A & S_B & | & S \\ M_A & M_B & | & M \end{pmatrix} \begin{pmatrix} S_A & S_B & | & S \\ M_A' & M_B' & | & M \end{pmatrix}^* \times \tilde{P}_\nu(M_A M_B; M_A' M_B')$$

where $\tilde{P}_\nu(M_A M_B; M_A' M_B')$ is a transition density between particular antisymmetrized products from (A.24). We want to eliminate the dependence on spin coupling by writing this density in the form of a spin-independent function multiplied by the matrix element of some suitable spin operator between spin functions with the eigenvalues indicated. What we can in fact show is that each transition density $\tilde{P}_\nu(M_A M_B; M_A' M_B')$ can be written in the form

$$\langle M_A' M_B' \mid (f_0^{(\nu)} + f_1^{(\nu)}(\mathbf{S}^A \cdot \mathbf{S}^B) + f_2^{(\nu)}(\mathbf{S}^A \cdot \mathbf{S}^B)^2 + \cdots) \mid M_A M_B \rangle$$

where $| M_A M_B \rangle$ is simply a product of spin eigenfunctions, $| M_A \rangle$ and $| M_B \rangle$, associated formally with the two systems. This means that, on adding the coupling coefficients and summing, each $\tilde{P}_\nu$ [and in particular the energy expression that follows, as in (A.13)] may be written as the expectation value in a resultant spin state $| SM \rangle$, of the same spin operator:

$$\tilde{P}_\nu = \langle SM \mid (f_0^{(\nu)} + f_1^{(\nu)}(\mathbf{S}^A \cdot \mathbf{S}^B) + f_2^{(\nu)}(\mathbf{S}^A \cdot \mathbf{S}^B)^2 + \cdots) \mid SM \rangle \tag{A.26}$$

The complete results have been given elsewhere (the first terms are listed by McWeeny and Yonezawa [*9*], the general terms of the expansions by Dacre and McWeeny [*10*]. Here, however, we give only the expressions for the normalization integral and the charge density, up to the single-interchange terms:

$$\tilde{P}_0 = 1 - \tfrac{1}{2} \int P_1^A(\mathbf{r}_1'; \mathbf{r}_1)\, P_1^B(\mathbf{r}_1; \mathbf{r}_1')\, d\mathbf{r}_1\, d\mathbf{r}_1' \\ - 2\langle SM \mid \mathbf{S}^A \cdot \mathbf{S}^B \mid SM \rangle \int D_S^A(\mathbf{r}_1'; \mathbf{r}_1)\, D_S^B(\mathbf{r}_1; \mathbf{r}_1')\, d\mathbf{r}_1\, d\mathbf{r}_1' \tag{A.27}$$

$$\tilde{P}_1 = P_1^A(\xi_1; \xi_1') \\ - \tfrac{1}{2} \Big\{ \int P_1^A(\mathbf{r}_1; \xi_1')\, P_1^B(\xi_1; \mathbf{r}_1)\, d\mathbf{r}_1 \\ + \int P_2^A(\mathbf{r}_2, \xi_1; \mathbf{r}_2', \xi_1')\, P_1^B(\mathbf{r}_2'; \mathbf{r}_2)\, d\mathbf{r}_2\, d\mathbf{r}_2' \Big\} \\ - 2\langle SM \mid \mathbf{S}^A \cdot \mathbf{S}^B \mid SM \rangle \Big\{ \int D_S^A(\mathbf{r}_1; \xi_1')\, D_S^B(\xi_1; \mathbf{r}_1)\, d\mathbf{r}_1 \\ + \int D_{SL}^A(\mathbf{r}_2, \xi_1; \mathbf{r}_2', \xi_1')\, D_S^B(\mathbf{r}_2'; \mathbf{r}_2)\, d\mathbf{r}_2\, d\mathbf{r}_2' \Big\} \tag{A.28}$$

where D_S and D_{SL} are the normalized densities introduced in

Lecture 4 (pp. 79, 89). The leading term in each case is the result appropriate to strong orthogonal groups. The nonorthogonality correction to the charge density is of particular interest because it depends on the spin densities of the two systems (the D_S terms) with a weight factor determined by the spin-coupling scheme. Since the $|SM\rangle$ are spin eigenfunctions with spins S_A and S_B coupled to a resultant S, M, the spin scalar product is equivalent to

$$\mathbf{S}^A \cdot \mathbf{S}^B = \tfrac{1}{2}\{\mathbf{S}^2 - \mathbf{S}^{A\,2} - \mathbf{S}^{B\,2}\}$$

and the expectation values in the various coupled states are thus

$$\langle SM \mid \mathbf{S}^A \cdot \mathbf{S}^B \mid SM\rangle = \tfrac{1}{2}\{S(S+1) - S_A(S_A+1) - S_B(S_B+1)\} \tag{A.29}$$

Since S runs in integer steps from $S_A + S_B$ down to $|S_A - S_B|$, it is clear that the scalar product may take both positive and negative values, the former corresponding to parallel and the latter to antiparallel coupling in the extreme cases.

To summarize: Each density function may be expanded in the form (A.26), and hence the whole energy expression for the composite system AB may be written as the expectation value of a spin Hamiltonian

$$\mathsf{H}_S = E_0 + E_1(\mathbf{S}^A \cdot \mathbf{S}^B) + E_2(\mathbf{S}^A \cdot \mathbf{S}^B)^2 + \cdots \tag{A.30}$$

in which the coefficients are determined by the electron density, spin density, and various coupling functions of the separate systems A and B. For any given spin coupling, it is then necessary only to insert the appropriate expectation values, as in (A.28), to obtain the corresponding energy expression. This is the result anticipated in Lecture 2 (p. 43).

Conclusion: Charge Density Changes in Colliding Systems

From the expressions (A.27) and (A.28) it is a simple matter to examine how the charge density changes when two systems collide and begin to interpenetrate. As we have seen, such effects depend strongly on the spin coupling as the approach commences and [in view of the simple electrostatic interpretation of Equation (4.13)] are of great importance in determining the nature of the interaction and the behavior of the potential energy surface. There is one potential energy surface for each coupling scheme, transitions between any two being forbidden until the separate systems overlap so much that their *internal* (Hund's rule) couplings are broken and more extensive configuration interaction must be admitted. It is therefore important to know, in a general way, what happens initially to the charge density in the region of overlap.

To answer this question we adopt a Hartree–Fock approximation to the wave function of each system, assuming that each possesses a closed-shell core together with an open shell of singly occupied orbitals with spins parallel-coupled (i.e., a Hund's rule ground state). Each system is then completely characterized by its electron density $P_1(\mathbf{r}_1; \mathbf{r}_1')$ and (normalized) spin density $D_S(\mathbf{r}_1; \mathbf{r}_1')$. Instead of the latter function it is convenient, in the Hartree–Fock approximation to use simply the difference of the up-spin and down-spin components in Equation (4.16), denoting this quantity by

$$Q_1(\mathbf{r}_1; \mathbf{r}_1') = 2SD_S(\mathbf{r}_1; \mathbf{r}_1') \tag{A.31}$$

In the same way, we replace D_{SL} in (A.28) by the function

$$Q_2(\mathbf{r}_1, \mathbf{r}_2; \mathbf{r}_1', \mathbf{r}_2') = 2SD_{SL}(\mathbf{r}_1, \mathbf{r}_2; \mathbf{r}_1', \mathbf{r}_2') \tag{A.32}$$

It is then a simple matter to show that

$$\begin{aligned} P_2(\mathbf{r}_1, \mathbf{r}_2; \mathbf{r}_1', \mathbf{r}_2') = {} & P_1(\mathbf{r}_1; \mathbf{r}_1') P_1(\mathbf{r}_2; \mathbf{r}_2') \\ & - \tfrac{1}{2} [P_1(\mathbf{r}_2; \mathbf{r}_1') P_1(\mathbf{r}_1; \mathbf{r}_2') + Q_1(\mathbf{r}_2; \mathbf{r}_1') Q_1(\mathbf{r}_1; \mathbf{r}_2')] \end{aligned} \tag{A.33}$$

$$\begin{aligned} Q_2(\mathbf{r}_1, \mathbf{r}_2; \mathbf{r}_1', \mathbf{r}_2') = {} & Q_1(\mathbf{r}_1; \mathbf{r}_1') P_1(\mathbf{r}_2; \mathbf{r}_2') \\ & - \tfrac{1}{2} [P_1(\mathbf{r}_2; \mathbf{r}_1') Q_1(\mathbf{r}_1; \mathbf{r}_2') + Q_1(\mathbf{r}_2; \mathbf{r}_1') P_1(\mathbf{r}_1; \mathbf{r}_2')] \end{aligned}$$

On inserting these expressions into (A.28) we obtain an expression for the unnormalized density matrix $\tilde{P}_1(\xi_1; \xi_1')$ in the form

$$\begin{aligned}\tilde{P}_1(\xi_1; \xi_1') = {} & [(1 - \tfrac{1}{2} M_P) P_1^A(\xi_1; \xi_1') - \tfrac{1}{2} P_1^B P_1^A(\xi_1; \xi_1') \\ & + \tfrac{1}{4} P_1^A P_1^B P_1^A(\xi_1; \xi_1') + \tfrac{1}{4} Q_1^A P_1^B Q_1^A(\xi_1; \xi_1')] \\ & - \tfrac{1}{2}\theta[M_Q P_1^A(\xi_1; \xi_1') + Q_1^B Q_1^A(\xi_1; \xi_1') \\ & - \tfrac{1}{2} P_1^A Q_1^B Q_1^A(\xi_1; \xi_1') - \tfrac{1}{2} Q_1^A Q_1^B P_1^A(\xi_1; \xi_1')] \\ & + \text{terms obtained by interchanging } A, B \end{aligned} \tag{A.34}$$

in which

$$\theta = \frac{\langle \mathbf{S}^A \cdot \mathbf{S}^B \rangle}{S_A S_B} = \frac{S(S+1) - S_A(S_A+1) - S_B(S_B+1)}{2 S_A S_B} \tag{A.35}$$

and the "products" are to be interpreted as, for instance,

$$P_1^A Q_1^B Q_1^A(\xi_1; \xi_1') = \int P_1^A(\xi_1; \mathbf{r}_1) Q_1^B(\mathbf{r}_1; \mathbf{r}_1') Q_1^A(\mathbf{r}_1'; \xi_1')\, d\mathbf{r}_1\, d\mathbf{r}_1' \tag{A.36}$$

(which is, in fact, the kernel representing the product of three integral operators). The quantities M_P and M_Q are defined by

$$\begin{aligned} M_P &= \int P_1^A(\mathbf{r}_1; \mathbf{r}_1') P_1^B(\mathbf{r}_1'; \mathbf{r}_1)\, d\mathbf{r}_1\, d\mathbf{r}_1' \\ M_Q &= \int Q_1^A(\mathbf{r}_1; \mathbf{r}_1') Q_1^B(\mathbf{r}_1'; \mathbf{r}_1)\, d\mathbf{r}_1\, d\mathbf{r}_1' \end{aligned} \tag{A.37}$$

and are essentially measures of the overlap between the charge and spin distributions of the two systems, respectively. With this notation, the normalizing denominator in the density matrix $P_1(\xi_1; \xi_1') = \tilde{P}_1(\xi_1; \xi_1')/\tilde{P}_0$ is given by

$$\tilde{P}_0 = 1 - \tfrac{1}{2} M_P - \tfrac{1}{2}\theta M_Q \tag{A.38}$$

It should be remembered that (A.34) and (A.38) neglect multiple-interchange terms; but it also follows easily that $P_1(\xi_1)$, even in this order, is exactly normalized in the sense

$$\int P_1(\xi_1)\, d\xi_1 = N_A + N_B \tag{A.39}$$

In other words, all the charge, amounting to $N_A + N_B$ electrons, is exactly accounted for. What we want to know is: How much of the charge stays distributed as it was in the separate molecules and how much goes into the region where they overlap?

To obtain a quantitative answer, let us imagine the density matrices expanded in terms of orthonormal sets of orbitals, one for each molecule $\{A_i\}$ and $\{B_j\}$. Thus

$$P_1^A(\mathbf{r}_1; \mathbf{r}_1') = \sum_i n_i^A A_i(\mathbf{r}_1)\, A_i^*(\mathbf{r}_1') \tag{A.40}$$

where the MO populations (n_i^A) are 2 and 1 for the closed and open shells, respectively. Then the quantities in (A.37) reduce at once to

$$M_P = \sum_{i,j} n_i^A n_j^B \,|\langle A_i \mid B_j\rangle|^2$$

$$M_Q = \sum_{\substack{i,j \\ \text{(open)}}} n_i^A n_j^B \,|\langle A_i \mid B_j\rangle|^2 \tag{A.41}$$

i.e., to weighted sums of the squares of ordinary overlap integrals between the orbitals of the two systems, the sum being confined to open-shell orbitals in the case of M_Q. The terms neglected in the present discussion involve fourth and higher powers of these overlap integrals, and even these are compensated for to some extent by the renormalization implicit in (A.39). It should be noted that both M_P and M_Q are essentially *positive*.

The charge density in the overlap region is described by the parts of $P_1(\xi_1; \xi_1')$ that refer jointly to A and B orbitals; and these are confined to the terms in (A.34) that involve only one A factor and one B factor. The total overlap *population*, in the sense of Lecture 4, is measured by the contribution of these terms to the charge-density integral (A.39). Let us denote this quantity (i.e., the total migration of charge from the separate systems into their region of overlap) by Δ. It then follows easily that

$$\Delta = -[(M_P + \theta M_Q)]/[(1 - \tfrac{1}{2} M_P - \tfrac{1}{2}\theta M_Q)] \qquad \text{(A.42)}$$

which depends very simply on the nature of the spin coupling between the two systems. The maximum and minimum values of θ, corresponding to parallel and antiparallel coupling, respectively, follow from (A. 35):

$$\theta_{\max} = 1, \qquad \theta_{\min} = -(1 + S_A)/S_A$$

where A denotes the system with greater spin ($S_A \geqslant S_B$). The extreme values of the overlap population (A.42) are thus (neglecting products of the small quantities M_P, M_Q)

$$\Delta_{\text{par}} = -(M_P + M_Q), \qquad \Delta_{\text{anti}} = (M_Q - M_P) + M_Q/S_A$$

For closed-shell systems, there is of course no spin-coupling problem, the θ terms in (A.42) are absent and we obtain $\Delta = -M_P$; charge is "pushed out" of the overlap region and there is repulsion between the two systems. Actual calculations [*8*] on the simple system of two hydrogen molecules indicate, in fact, an approximately linear relationship between repulsion energy and total overlap population. For open-shell systems, on the other hand, the singly occupied (valence) orbitals are the first to overlap significantly; neglecting closed-shell overlap, we then obtain initial estimates

$$\Delta_{\text{par}} \simeq -2M_Q, \qquad \Delta_{\text{anti}} \simeq M_Q/S_A$$

Thus, while parallel-coupled open shells are twice as effective as closed shells in producing repulsion, antiparallel coupling actually produces a bonding effect—electron charge migrating into the overlap region.

The simple conclusion is that systems approaching with spins parallel-coupled repel one another, their charge distributions being pushed apart like rubber balls; but systems with spins antiparallel-coupled attract one another, charge piling up in the region of interpenetation. We are now dealing with arbitrary many-electron systems, described by highly accurate wave functions; but the remarkable fact is that the Heitler–London idea—that interactions may be described *formally* in terms of a coupling of spins—still survives intact.

REFERENCES

1. Moffitt, W., *Proc. Roy. Soc. (London) Ser. A* **210**, 224 (1951).
2. Parr, R. G., Ellison, F. O., and Lykos, P. G., *J. Chem. Phys.* **24**, 1106 (1956).
3. McWeeny, R., and Mizuno, Y., *Proc. Roy. Soc. (London) Ser. A* **259**, 554 (1961).
4. McWeeny, R., and Sutcliffe, B. T., *Proc. Roy. Soc. (London) Ser. A* **273**, 103 (1963).
5. McWeeny, R., *Proc. Roy. Soc. (London) Ser. A* **253**, 242 (1959).
6. London, F., *Z. Physik* **63**, 245 (1930).
7. Longuet-Higgins, H. C., *Proc. Roy. Soc. (London) Ser. A* **235**, 537 (1956).
8. Magnasco, V., Musso, G. F., and McWeeny, R., *J. Chem. Phys.* **47**, 4617 (1967).
9. McWeeny, R., and Yonezawa, F., *J. Chem. Phys.* **43**, S120 (1965).
10. Dacre, P. D. and McWeeny, R., *Proc. Roy. Soc. (London) Ser. A*, to be submitted (1970).

AUTHOR INDEX

Numbers in parentheses are reference numbers and indicate that an authors's work is referred to although his name is not cited in the text. Numbers in italics show the page on which the complete reference is listed.

SUBJECT INDEX